Meetings with Remarkable Mushrooms

Meetings with Remarkable Mushrooms

Forays with Fungi across Hemispheres

Alison Pouliot

The University of Chicago Press
CHICAGO AND LONDON

The University of Chicago Press, Chicago 60637
The University of Chicago Press, Ltd., London
© 2023 by Alison Pouliot
Published 2023
Printed in the United States of America

32 31 30 29 28 27 26 25 24 23   1 2 3 4 5

ISBN-13: 978-0-226-82963-0 (cloth)
ISBN-13: 978-0-226-82964-7 (e-book)
DOI: https://doi.org/10.7208/chicago/9780226829647.001.0001

First published in Australia by NewSouth, an imprint of UNSW Press Ltd.

Library of Congress Cataloging-in-Publication Data

Names: Pouliot, Alison, author.
Title: Meetings with remarkable mushrooms : forays with fungi across hemispheres / Alison Pouliot.
Description: Chicago : The University of Chicago Press, 2023. | Includes bibliographical references and index.
Identifiers: LCCN 2022059969 | ISBN 9780226829630 (cloth) | ISBN 9780226829647 (ebook)
Subjects: LCSH: Mushrooms. | Fungi.
Classification: LCC QK617 .P668 2023 | DDC 579.6—dc23/eng/20230130
LC record available at https://lccn.loc.gov/2022059969

For my father

CONTENTS

A Note on
Fungal Terminology

Finding a common language to talk about fungi can be tricky, especially when they're known by different names. A fungus species often has several vernacular or common names, and one common name can also refer to several species. In North America, for example, "meadow mushroom" or "pink bottom" is the common name for an edible mushroom species keenly sought by foragers. In Australia it is known as a "field mushroom." Others in the English-speaking world refer to it by the more generic French name for fungi, "champignon." The common names used in this book are those used by the people in the places where a particular fungus was found.

Each species has only one scientific name, although that can change as more is learned about relationships between species and their classifications are rejigged. Scientific names are binomials, meaning they have two parts: the first is the genus and the second is the species. The abovementioned common names, for example, all refer to the fungus *Agaricus campestris*.

Common names and scientific names serve different needs. While scientific names allow scientists to talk about the same species across languages and cultures, common names are often more familiar and accessible to the lay reader. For ease of reading, I have used common names in the text of this book, and included scientific names in the listing at the end. That said, some scientific names also appear in the text, as many fungi are yet to be given common names. Scientific names are also used where needed to avoid confusion. When a genus name appears more than once in nearby text it is usually abbreviated. For example, *Phellinus rimosus* is written in full on first mention and abbreviated to *P. rimosus* on subsequent mentions. Further species within an aforementioned genus also appear in the abbreviated form, e.g. *P. robustus*.

As with all specialist fields, mycology—the scientific study of fungi—has its own vocab. While technical language can clarify, it can also obscure. I try to balance precision with accessibility and define terms as I go, but I have also included them in a glossary at the end of the book.

To refer to the characters or habits of fungi, words are often borrowed from plants and animals. The reproductive receptacles of fungi are often called "fruiting bodies." However, "fruit" is a botanical term that describes a plant structure that contains seeds. Fungi are not plants and don't have seeds, but spores, hence I adopt the increasingly used term "sporing bodies."

Like fungi, places are commonly known by several names. Australia once had more than 250 Indigenous languages and many more dialects. European colonization saw the silencing of many of those languages, and the renaming of places and geographical features. Throughout this book, I have used Indigenous place names where they are known, followed by the colonial name on first mention.

Reinstating the Indigenous names of places is part of recognizing Australia's Indigenous history. It also adds to the story of fungi. Early colonial collectors of fungi sent specimens to England for identification, and most of them are still there, housed in the bowels of the fungarium (the fungal portion of the herbarium) at Kew Gardens. While the fungi may never return, we can at least return the names to the Country from where they were taken. I adopt the capitalized word "Country" in the context of it being more than simply a place or territory, to represent the more relational Indigenous idea of belonging and deep connection to land.

1

STIRRINGS IN THE SUBTERRAIN

It was raining in Whitby. That was hardly unusual, and I should have been pleased. Everyone knows rain brings mushrooms. Westerly winds swept over the North York Moors, delivering showers in squalls and spurts along England's Yorkshire coast. I was there for an international congress on fungal conservation and the dampness boded well for fruitful field trips. But I was trying to hitch a ride to the congress dinner and was already late, and the rain ran cold down the back of my neck.

A vehicle appeared, slowed almost to a standstill then sped off, spraying me with water. The road returned to darkness. Gulls mewed in the distance as another vehicle approached, blinding me with its headlights, but it slowed and stopped. I ran toward it. It was only when the tailgate flipped down and the back window flipped up that I saw it was a hearse.

"Well get in then!" barked a voice in a North Yorkshire accent. I probably should have hesitated, but I didn't. I leapt in and cracked my knee on something hard. It was a coffin. A coffin in a hearse shouldn't seem odd. But the three goths sitting bolt upright inside it drinking champagne did. They eyed me suspiciously as one passed me a glass. He filled it to the brim and champagne overflowed down my sleeve. "So where'd you be goin' this fine evenin'?" he asked. My knee throbbed and I could feel my hair stuck to the sides of my face as I shook out my sleeve. "To a congress dinner on the quay," I replied, then added, "a congress on fungi!" He squinted and pursed his blackened lips. No one spoke. The windscreen wipers flapped louder.

You'd have thought that with our common interest in the subterrain, I'd landed with kin, but the other two goths glanced sideways, and one raised an eyebrow. It seemed the fungus congress and the Whitby Goth Weekend had been programmed simultaneously and we each thought the other the more strange. But after a prolonged silence, the questions about fungi came thick and fast until the driver cranked up the stereo and The Sisters of Mercy saved me from further interrogation.

As we turned onto the quay, I caught sight of the ruined Whitby Abbey perched high on the headland overlooking the North Sea. Pedestrians dashed across the wet road. "Here! Stop here, please!" I yelled to the driver and he pulled into the curb. I handed back my glass and wished the goths well. They nodded in unison as I climbed out of the hearse. Back in the rain, I paused for a moment to gather myself, then headed toward the lights of the restaurant, certain that my foray into fungal realms would be every bit as thrilling as the ride.

The International Society for Fungal Conservation Congress drew a motley band of conservationists, fungus enthusiasts, and mycologists—scientists who study fungi—from the forest and the laboratory to David Minter's hometown of Whitby. David is a mycologist and the mastermind of the society. A natural-born showman, he's good at holding court, convincing anyone who might not yet be convinced that fungi should be at the heart of biodiversity conservation, not the periphery.

David has been fighting for fungi and their recognition for a long time. Fungi seldom feature in conservation because they seldom feature in our ideas about what that thing out the window—nature, biodiversity, the environment, whatever you want to call it—actually is. But by only considering above-ground ecologies of plants and animals, what if we were failing to protect the diversity of fungi below ground? What if this oversight meant a slow, unseen unraveling of the very foundation that enables all aboveground life to flourish? Even the scientists who study organisms other than fungi and the conservationists who rally for them are usually largely unaware of the need to conserve fungi. However, given most of those organisms are intertwined with fungi in some way, including fungi in conservation makes good sense.

The society is the first in the world to explicitly and exclusively focus on protecting fungi. The congress delegates were there to tease out why fungi have been overlooked and what's needed to bring them into public awareness and onto conservation agendas. I thought about the goths and how their subculture, like mycology, is often seen as fringe, offbeat, even dubious. Yet they too were challenged by the idea of a congress on fungi. What is it about fungi that presents such a challenge to people?

Kingdom Fungi is one of conundrums. For some people, fungi are unnerving. Perplexing. Enigmatic. Being ephemeral, they rouse suspicions and suggest associations with the supernatural. But mostly, fungi are overlooked, forgotten. Fungi may have finally found their own kingdom back in 1969, but many people still struggle to explain what a fungus is or what it does. And fungi continue to unsettle the ways we think about the natural world. They do things we don't understand. They're unruly. They may enter periods of dormancy or respond to environmental cues of which we're unaware.

Yet fungi captivate. With their seemingly inexplicable presence, their curious forms and potent apothecaries, they've stirred imaginations across the centuries, challenging us to open our minds and expand our thinking. Fungi disrupt our frameworks for ordering and understanding nature. They unravel old assumptions about how forests function. With their extraordinary beauty and strangeness, but also their utility, fungi are luring new fans.

—

Just over a decade later and half a world away, I stood in the main street of the south-east Australian town of Euroa. A semitrailer was parked across the traffic circle, blocking Binney Street to cars. Something was brewing. The curious lives of fungi Down Under were infiltrating the minds and imaginations of an ever-growing band of mycophiles—people who are enamored with fungi—and an upsurge of interest had hit an all-time high. I sensed we were approaching peak fungus.

Led by artist Penny Algar, a group of local conservationists and fungus enthusiasts climbed onto the tray of the semi to launch the third biennial Strathbogie Festival of Fungi. I skirted the gathering crowd of people and saw a flash of purple light as the old redbrick post office exploded in a gyre of fungus spores. The spores drifted onto the wide-eyed faces of children below and slowly floated down Binney Street. Glimmering like distant stars, the digital spores of light beamed from a projector, powered by a man pedaling a bicycle. I stood mesmerized as shopfronts and walls, trees and passing dogs all lit up in a captivating visual feast of illuminated fungi.

The festival had lured locals and visitors from the warmth of their lounge rooms into the bitter winter night.

Clad in beanies and woolen scarves, they stared transfixed as the unfolding fungal lightshow flashed and flickered around them. Mushrooms loomed from shop windows. A child scrabbled on the ground trying to grasp the elusive fungal threads beneath her feet. Along the creek, an old river red gum transformed into a shifting canvas of psychedelic fungi.

Why were all these people so excited? It's a pretty sure bet that if you project some moving images onto old buildings in an empty street, people will notice. But there's a growing public penchant for fungi that suggests we are in something of a fungal awakening. An emerging league of mycophiles is delving into the many dimensions of Kingdom Fungi. Some are tapping into fungi to help restore stressed ecologies. Some grow mushrooms on their kitchen tables, resisting the monopolies of industrialized agriculture. Others head to the forest to forage for fungi or seek deeper connections with nature. Writers and artists weave fungi into their work. Bioengineers and myco-entrepreneurs explore possibilities for fungal alternatives to building materials and plastics. Fungi steadily edge their way from the margins to the mainstream. The digital spores of Strathbogie's festival illuminated a usually hidden realm of life. As people came to meet with mushrooms, the misgivings of mycophobia—the fear of fungi—that grip the English-speaking world faded into the night.

Growing interest in fungi plays out elsewhere in the world, too. And often in strange ways. The previous year, in the northern hemisphere, I'd pulled into the car park of the North Seattle College and watched transfixed as a giant morel mushroom—or rather, a human being disguised as one—attempted to extract itself from the cabin of a pickup truck. Fungi were popularized in the United States in the 1960s through psilocybin mushrooms or magic mushrooms, but public interest in fungi has grown more eccentric than the psychedelic. With much maneuvering and grunting, the morel finally unshackled itself, sprouted two feet and hit the ground with a groan. Adjusting its handbag, it then waddled down the path ahead of me. A dog scuttled sideways, tail between its legs as a toddler released a long howling wail. Undeterred, the morel wobbled precariously onward.

It seemed the morel and I were both there for the same reason, which was to attend the Puget Sound Mycological Society Annual Wild Mushroom Show. I was evidently under-dressed. The show attracts people from a range of fungal fetishes to learn of the latest mycological revelations, poke and prod at displayed fungi, and perhaps ponder the human bent for dressing as mushrooms. Mycophiles converge from across the country. Those sufficiently skilled at distinguishing delectable fungi from those that are deadly savor the Annual Survivors' Banquet. Beyond the Mushroom Show, the society supports amateur and professional mycology in America's Pacific Northwest with an active program of research and education.

At times, the burgeoning interest in fungi pops up in surprising places. The WOMADelaide festival, which brings together musicians, dancers, and artists, introduced a forum called "The Planet Talks" to discuss and debate science and nature themes. At the 2019 festival, I met with mycologist Brian Pickles and bioengineer Gavin McIntyre, along with panel host Robyn Williams, in Adelaide's Botanic Park to talk fungi. Robyn has hosted *The Science Show* on ABC Radio since 1975, making it one of the longest running science programs in the world. After the gig I asked him what he thought was inspiring the "fungal turn." Robyn spoke of growing public knowledge of soil, the connections between fungi and plants, and how "many of us have now seen the thrilling complexity under all those trees we once thought were lone statues."

This shift in thinking from lone statues to thrilling complexity reflects the public recognition of fungi as more than mushrooms. The groundswell of interest is part of a bigger ecological turn, sparked perhaps by concern about climate change, the demise of species, and the need to avert our precarious trajectory. It is part of a research revolution that is shifting the way we think not only about fungi and the natural world, but the ways in which knowledge develops. Like mushrooms themselves, the welling of interest swells steadily from the ground.

From high-tech labs, abandoned garages, and everything in between, innovations and applications of myco-technology are emerging at a rapid rate. Many hold promise for developing fungal alternatives for remediating damaged environments. The great challenge is to scale them up to a useful or meaningful level. However, we are unlikely to find solutions to the environmental issues created by humans in a technological fungal fix. That requires a change in thinking.

Remediating our relationship with the natural world could be a first step toward using fungi to remediate environments. Fungi won't save the world, but they could they offer insights into more sensitive ways of being in the world.

—

I've got to know fungi mostly by spending time with them, at their place in the forest, as a naturalist and an ecologist, a conservationist and photographer. Most fungi appear in the fall and usually only for a short time. I quickly realized that if I could track fungi across hemispheres and benefit from two falls each year, I could maximize my time with them. So over the last twenty years, I've spent six months in Australia or elsewhere in the south, and six months in the northern hemisphere, mostly in Europe. I've ventured into fungal realms with mycologists and ecologists; field naturalists, foragers and film makers; farmers and First Nations communities. All have shared their take on fungi, each in their own time. Their collective perspectives reveal not only the complexity of fungi, but the many lenses through which they are understood. Each experience has enriched my own understanding of fungi.

I've been getting to know fungi for a while. As a child, I was enchanted by the Australian bush. Entering a forest sparked an immediate sense of excitement. My heart thumped faster, but it wasn't a rowdy kind of excitement— more an intense welling of wonder, a sense of ease of slipping into rhythm with the forest. Being in nature sharpened my senses and fueled my imagination. It felt like it was where I was meant to be.

With the gift of a loupe or handheld magnifying glass at the age of seven, I discovered hidden marvels—hairy caterpillars, the filigree venation of leaves, compound eyes of flies. I now had a means to visualize these mysterious microcosms. But fungi held another kind of allure. It was the delicate underbellies of mushrooms in all their configurations, and the creatures that inhabited them, that had me spellbound. They were whimsically strange. Startling. Incongruous and extraordinary. They grew in unexpected places and appeared when no one was watching. I sensed there was more going on and I wanted to know what they were doing. My curiosity soared. I was learning how to see. Nearly five decades later, I treasure that magnifier. Peering through it today is no less enthralling.

Observing fungi through lenses from an early age was good training to be a scientist, but my fascination with fungi began with their aesthetics; with noticing their beautiful and bizarre forms. I marveled at the delicacy and exquisite perfection of umbrella-shaped mushrooms, but it was the other more eccentric fungal forms like corals, clubs, and lattice balls that captivated me. Over time, I peered through lenses attached to cameras and microscopes. Science and aesthetics were inseparable. As I grew older, this close-up view alerted me to disturbing environmental changes, often slow and insidious, undetected and unobserved, at larger scales: the toxin-induced deformities of midge larva mouthparts; the disturbing presence of plastic particles in soil samples; malformations on mushrooms. These lenses provided other views of life that have taken me on a lifetime search for answers.

Fungi have always made most sense to me in the dynamic context of their environments. These stories about fungi are therefore entwined in those about soils, trees, and forests, and the creatures that inhabit them. Fungi simply provide an entry point. I hope to bring you with me into the forest, where I'm writing this, in case you're interested, with a 3B pencil. It feels nice in my hand and slides easily across the page of my old oilskin notebook. Stories of fungi are smudged with raindrops, spattered with mosquito legs and fungus spores. Here in the forest, I can reach out and touch a fungus and describe whether it feels like flesh or felt, rubber or suede. They're astonishingly tactile. Being in the forest is a lifelong process of noticing minute details and subtle changes, regularities and peculiarities, and wondering what each might mean.

—

It's unusually still on the southern slopes of the mountain in Australia's southernmost state of Tasmania. Known as Kunanyi or Mount Wellington, to the locals in Hobart it is simply "the mountain." I crouch in the undergrowth. With frozen fingers I adjust my camera lens and a huddle of tiny red mushrooms comes into focus. Known as ruby bonnets, they are blood red, perfect, and I'm probably the only human who'll witness their fleeting existence.

Mists waft gently around me and slowly drizzle water through the fronds of tree ferns. Lichens wrap around twisted blackwood limbs like ornate scarves. They're rigged with the overnight industry of spiders, their webs jeweled with droplets of condensed fog. In the gloom of the forest floor, fallen branches are sheathed with fungal stripes of yellow and purple. But beneath the colorful surface, the fallen litter is alive with the clandestine workings of fungi.

As I slip through wet fronds, a striking strawberry bracket fungus looms from the trunk of an old myrtle beech. I run my fingers over the contorted fungal lobes. They're like velour and feel as lovely as they look. Despite this fungus being conspicuous, it's the first time I've seen it here. The

British-born Tasmanian botanist Leonard Rodway might have been similarly spellbound; he named it *pulcherrimus*, meaning "most beautiful."

People understand their surroundings and the species that inhabit them in different ways. From her house in Hobart's Sandy Bay, my grandmother had sweeping views of the mountain. Each morning she looked toward it to get an impression of what the day had in store. She noticed changes in wind direction, the nature of the clouds, or perhaps a dusting of snow on the summit. She didn't need a weather app. The mountain was her gauge.

Traveling with her at walking pace—my grandmother didn't drive a car—allowed more time for noticing, for stopping and greeting friends along the way. While she talked, I watched the birds and the lizards warming themselves on Hobart's sandstone rock walls. Understanding fungi also begins with noticing. People who live close to the land are more likely to be aware of the ways of mushrooms— when and where they appear, and especially if they don't. Keen noses sniff out clues that reveal their cryptic presence. Through trial and error, they know which fungi harm and which may heal. Lives are lived in reciprocal relationships with the land. They know to let fungi be.

Being immersed in the sensorial world of the forest is about intimately understanding the local. Ecologies can vary greatly over small areas and short spaces of time. Local observations of subtle shifts and changes forge a rich and nuanced understanding of the forest. It's more finely honed with particularities than the blunt assumptions with which we try to make sense of a globalized world. To conserve an environment means knowing about it, but it usually begins with caring about it. I've worked with conservation-minded people on the ground across the world and have observed that caring almost always happens locally. Their connections to place, curiosity, and pragmatism come together as practical wisdom, as an environmental ethic. Understanding something emotionally is as important as being able to explain it.

Although science is by nature objective, scientific exploration often starts with a feeling. We notice a fungus, we become curious about it, we follow a hunch or intuition, we ask a question, and we go about trying to answer that question. Is there a story to tell? Like intuition, serendipity has always played a role in science. Happenstance is the unwritten part of the research process and often the most revelatory. Many interesting insights from my field research have arisen from unexpected encounters and accidental discoveries. The overgrown track that sent me down another path. The superb lyrebird scratching for invertebrates in leaf litter that flung truffles my way. A fungus casually observed while searching for another.

What we understand intuitively is sometimes seen as removed from science. But local knowledge gathered over time can collectively be richer than the templates and protocols with which humans attempt to manage ecosystems. Intuition develops from experience and accumulated knowledge, and it lies at the heart of the scientific process.

—

While mycologists and biotechnologists reveal the wonders and potential of fungi, they are relative newcomers to the field. Many European, South American, Asian, and African cultures have long used fungi as food, medicine, and psychotropics, while those in the Anglosphere have been a little slower to catch on. Foraging for edible fungi is likely to be as old as the history of food gathering itself and is a scientific process of trial and error. The oldest evidence for the use of fungi as food belongs to the Red Lady, a Magdalenian woman of the late Upper Paleolithic who is thought to have lived 18,700 years ago, according to radiocarbon dating of her burial site in northern Spain. Examination of the dental calculus from her teeth suggest mushrooms were part of her diet. But given the more than 60,000-year-old history of Aboriginal people in Australia, their use of fungi could be tens of thousands of years older.

Today, some in the English-speaking world have a great disdain for fungi, and many fear them, earning them the reputation of being mycophobic. Most simply have no interest in them as food or otherwise. So why are the Anglo cultures so fearful? One theory is that people lost connection to their surrounding environments as the public fields and forests where they foraged became privatized. They not only lost access to the land and its mushrooms, but the knowledge that accompanied foraging. The revolutionizing of agriculture by private landholders saw crops planted, harvested, and sold, and foraging was only for the nomadic and the poor. Contempt for the forager likely spread throughout the British Empire. Part of the fungal awakening is the effort to retrieve lost knowledge and traditions of foraging for wild fungi.

Fungi began to find favor again in the English-speaking world during the 1970s and 1980s, though the oldest mycological club of the hundred or so in the United States, in Boston, was founded in 1895. Public interest in fungi has grown rapidly over the last 50 years, especially in the last two decades. Across the Atlantic Ocean, the British Mycological Society was founded a year later than its Boston counterpart, but the groundswell of interest in fungi in the United Kingdom similarly surged in the 1980s.

In Australia, although scientific interest in fungi had always been on the agenda of the Field Naturalists Club of Victoria (founded in 1880), fungi found new followers with the establishment of Fungimap in 1996. The first fungus-mapping scheme in the southern hemisphere, Fungimap united a small but keen coterie of mostly field naturalists to work out which fungi grew where in Australia. Citizen scientists now help mycologists put dots on maps by recording the fungi in their surroundings. Other not-for-profit community groups are springing up, waving the flag for fungi and increasing public awareness of their importance. Growing interest in foraging and cultivating fungi, and the melding of old knowledge with new discoveries, are all part of the fungal awakening.

—

This book brings together some of my experiences and stories about fungi. They came about through journeys to meet both fungi and their followers: sometimes a bettong hot on the trail of a truffle, at other times a mycologist jammed under a log. The stories I tell are founded in science but told by more than data. They are observations and anecdotes, discoveries and revelations, lived experiences of fungi, stories about life felt as much as known. I hope this book stirs the senses and imagination, as much as it informs.

These stories sit at the nexus of modern mycology and the animate worlds of indigenous peoples, the creative expression of bio-artists and the toil of those working in the dirt to restore fungal ecologies. It gathers diverse voices and strands of knowledge to consider fungi from different perspectives. Curiosity drives them all.

Fungi have lured me to diverse terrains. *Underground Lovers* charts a fungal adventure from the towering Sitka spruce forests of the Pacific Northwest of the United States to the ancient oaks on the Italo-Swiss border and the lichen-scapes of Iceland. In the southern hemisphere we travel from Australia's Top End to Tasmania in the far south, and "across the ditch" to some of the world's wettest temperate rainforests in New Zealand. Although the trees and fungi differ in how they look, feel, and smell, don't be fooled. Head into the subterrain and underground lovers do much the same thing world over—they find partners, share, squabble, transform, and hold forests together. They do so with astonishing complexity and dynamism that is at the core of the fungus-forest story. Let's head into the forest.

2
LIFE IN THE
MYCOSPHERE

I never intended to go mushrooming in a V8 Dodge Charger Hellcat. It was one of those mix-ups at the car-hire place, and the Hellcat was the last vehicle available. I was told in no uncertain terms: "Listen, ma'am, you can take the car or walk." It wasn't the field vehicle I'd imagined, being low to the ground with the clearance of a porcupine. But I climbed into the supercharged machine and watched the Seattle skyline rapidly disappear in the rear-view mirror.

I'd found myself in the United States after an invitation from mycologist Steve Trudell to present a series of talks and forays in the fungus-filled forests of the Pacific Northwest. At the first meeting with the Snohomish County Mycological Society, I slid the Hellcat in among the Subarus and other sensible cars. When it roused more questions than mushrooms, I decided that from then on I would park it around the corner. Fungi returned to the conversation, and I came to appreciate how quickly the 'Cat could deliver me from the asphalt to the forest.

Some of the most impressive of those forests grow on the Olympic Peninsula in Washington State, where mycologist Lee Whitford lives. She invited me to explore her "backyard"— more than 1 million acres of temperate rainforest, glaciated mountain ranges, alpine lakes, and rugged coastline of the Olympic National Park. And if that wasn't enough, there was the lure of the 1,400 species of fungi known to live there. "We'll go from valleys filled with big-leaf maples and red alders up to the subalpine areas, with mountain hemlock and silver firs, to discover all their different fungi," Lee said, temptingly. It took her a fraction of a second to convince me. We set off before dawn (in Lee's sensible Subaru) through swirling fog and squalls of rain. It was perfect weather for fungi. Nothing could deter us.

A couple of hours into the drive, Lee slowed, peered into the forest, then veered off the road down an unmarked track. Deep in a valley carved by the Sol Duc River, she took me to one of her secret fungus-hunting haunts. It wasn't secret in the way of a forager's secret spot, which has the prospect of edible mushrooms, but because the forest was old and rich with rare and unusual fungi.

Forests are often deemed "old growth" by the age of their trees. By nature long-lived, some trees reach many hundreds of years. But the subterranean lives of some fungi rival their woody friends. Older forests generally harbor a greater diversity of fungi than younger ones, and therefore make new discoveries more likely. "I love seeing these forests through other people's eyes," said Lee. "Everyone's pattern recognition or fungus search image is a little different, so we always find something new."

Stepping into the temperate rainforests of the Olympic Peninsula is like entering a lost primeval world. The Strait of Juan de Fuca and the Pacific Ocean deliver dripping fog along with 140 inches of annual rainfall. These forests are all wetness and exuberance. It takes a while to register the great palette of greens and the mesmerizing quality of light filtering between them. Evergreen conifers dominate. Spires of Sitka spruce disappear into low cloud. Western red cedar and Douglas fir stand shoulder to shoulder.

The majesty of these giant elders captures attention, but the character of the forest is expressed in its epiphytes—organisms that grow on plants but are not parasitic. With little space on the crowded forest floor, they take advantage of any trunk or branch with a spare spot to squeeze in. Tapestries of mosses and liverworts slow the movement of water, gently dripping it to the trees' roots and topping up the soil reservoir. Licorice ferns clamber up ancient trunks, between lungworts and cat-tail mosses. Light is limited in the understory. For most plants, it helps to get a leg up into the light. But many epiphytes cope with low light and colonize the larger trunks where there's more wiggle room. They wrangle the delicate balancing act of finding space, seeking light, and staying low in the dampness.

It takes a while to get your head around the profusion of life jostling together in these forests. This is high-density living at its most extreme. It's hard for a falling seed to reach the soil through the dense understory. Rather, many seeds germinate on fallen trees, or nursery logs, that nourish the seedlings with moisture and nutrients as they slowly decay.

We pause by an old fallen western hemlock. Fungi work away within its depths, slowly transforming it to soil. Translucent arcs of a fungus known as angel wings cling to its undersides. Splashes of color punctuate the shades of green as the yellow funnels of winter chanterelles push through the moss on the forest floor. Giant banana slugs cruise in between, sampling the smorgasbord of fungi on offer. I reach through the ferns to stroke the velvety caps of violet corts. They feel more animal than fungal.

Then Lee calls out: "Ooooh, have you ever seen these?" I scramble through the undergrowth to where she is lying awkwardly between logs. I smile, knowing from my own desire to photograph fungi that it can demand perverse, gymnastic acts of extreme discomfort. Lee gently parts the moss to reveal two unfamiliar purplish-brown mushrooms. Their glutinous caps glisten. "Ah, but *this* is the really exciting bit," she says, reaching into her basket for her pocketknife. Lee gently excavates around one of the mushrooms and, as she draws it out of the ground, its strange and twisted taproot-like stipe (stem) is revealed. "It's a *Phaeocollybia*, or a rootshank. They're fairly common here on the peninsula, unlike elsewhere. It's a robust thing and can be four to five inches tall, but when you dig down, the stem continues another six inches!" The demand for space and resources continues below the forest floor. The rootshank's long stipe suggests its mycelium lives deeper in the soil than that of most fungi. Finding a nook in the forest might mean that they, too, need to be gymnastically adept.

Lee knows her local fungi, and I'm enraptured by forms and species new to me and by her erudite and gentle explanations. We spy an odd black lump on the stump of a Douglas fir. It's an Oregon black truffle with a suspicious track of teeth marks exposing its gray innards. It seems we have interrupted someone's breakfast, a southern red-backed vole perhaps, abandoning its fare in a hasty escape. Like fungi, the voles inhabit the rich leaf litter–soil interface, where they consume truffles and disperse their spores.

We clamber over a log and with each step more fungi loom through the green. The contorted forms of the western elfin saddle, each of them unique, compete for the prize of most eccentric mushroom. "What kind of evolutionary pressures throw up such comic configurations?" asks Lee rhetorically. "It's extraordinary how all these shapes, colors, scents, and textures seemingly emerge from nothing." We both know they emanate from the hidden network of mycelium, but I agree that at times it feels more miraculous than that.

Lee and I gradually fall into an easy introspective silence, enjoying each other's company while entranced by our own discoveries, absorbing the mood of the forest. As I struggle to find a spot to step among the profusion of life on the forest floor, there's a sense that something is driving it all, holding it all together, unseen beneath the surface. It feels as though the forest literally vibrates with the lively activity of subterranean industry.

Heading Below Ground

For many of us, fungi are puzzling because we make sense of nature based on our understanding of animals and plants. That's what was taught at school, at least in Australia. Fungi are often defined not by what they are, but by what they are not. As more familiar organisms, animals and plants are usually the reference point against which fungi are described: "a plant without chlorophyll," "a disease of frogs." But fungi are very different organisms from plants and animals in the way they are built and operate. Although they have simple body plans, they have obscure structures and unique traits and behaviors. To grasp what fungi are and do in ecosystems, it helps to get a sense of how they grow and feed, the nature of their interactions with plants, and their pivotal role in transforming environments, recycling organic matter and creating soil.

Once it was mushrooms that attracted attention. They still do, but growing interest in mycelium and the notion of subterranean networks of fungi is changing not only how we understand fungi and forests, but life itself. Fungi provide a fundamental foundation to the forest, and are a key to understanding how forests work. So, for a moment, let's zoom in: from forest, to mushroom, to mycelium, to hypha to spore.

The life of a fungus begins with a spore. If the spore happens to land somewhere favorable to germination, a gossamer-thin cylindrical cell known as a hypha emerges. An individual hypha is microscopic, but we can see them en masse as a fungus mycelium. If you gently part the leaves on the forest floor or look beneath a log, you're likely to see web-like fungus mycelium.

Fungi are unusual. Unlike other multicellular organisms that develop by growing more cells in layers, fungal hyphae grow by elongating themselves. To do so, the tip of each hypha needs new material to insert into its cell wall. This material arrives within the cell in a flow of bladder-like sacs that contain the necessary building materials and enzymes to enable the extension. It doesn't happen by magic, although the process is perplexing. The choreographer that directs it is a curious subcellular structure or organelle called the Spitzenkörper, meaning "apical (pointed) bodies" in German. It's one of several features that make fungi entirely different organisms from animals and plants.

Hyphae have two special talents that enable them to form a mycelium. The first is to branch. One hypha branches and forms two hyphae. However, their more unusual talent is that hyphae can anastomose, or fuse. Few other organisms do this. By branching and fusing, hyphae grow into the complex interactive network of the fungus mycelium. If two hyphae from the same mycelium meet, they bridge and fuse, connecting different parts of the mycelial network. To do this, one hypha literally dismantles part of the outer wall of the other. It is a delicate process. A two-way flow of cytoplasm—cell fluid containing ions, molecules, organelles, and other vital goodies—moves between them. Cell nuclei are also transferred. This helps the mycelium self-regulate and allows for the exchange of genetic information. A cell-wall "mortar" is then applied to join the two.

Furnished with a great swag of enzymes, and cell walls made of chitin—a structural compound that gives them hardness, which is also found in arthropod exoskeletons—hyphae are well equipped to explore their surroundings in search of food. The ability to branch allows them to grow in every direction, their feeding hyphal tips probing the soil seeking organic matter that provides them with nourishment. When part of the mycelium encounters something nutritious, it secretes enzymes from the tips of the hyphae, breaking it down into smaller absorbable molecules. The rest of the mycelium responds to the discovery by concentrating its growth toward it. While the shape and form of a mycelium varies between species, it is in part determined by the location of food.

A mycelium can technically expand infinitely. As long as there is a continuous food supply and no one excavates its home, builds a road on top of it or otherwise disturbs it, the mycelium continues to grow beneath the soil. If you've ever flown over the braided waterways and mudflats of the Sundarbans Delta in the Bay of Bengal or the meandering channels of Australia's Kimberley coast, you can probably imagine what mycelium looks like. The mystery of how it communicates across its expanse is slowly becoming understood, and chemical and electrical signaling are known to play a part. My sense of the "vibrating forest," whether real or imagined, exemplifies the vast wefts of the mycelial tangle that are more than merely static architecture but dynamic processes of flows and feedback loops.

This vibrant mycelial scaffold underpins the forest. I try to comprehend the flourishing life forms around me that are exponentially mirrored beneath my feet. Compared to the size of the velvety violet cort that could fit easily in the palm of my hand, the subterranean mycelial networks are much more expansive. Meshworks of mycelia bind soil particles. They also create spaces in between. By aerating soil, they make it habitable for other organisms that have evolved to live underground and that contribute to the recycling and liveliness of the soil. The mycelial architecture allows water to gently trickle down to the soil's deeper horizons. Without scaffolds of mycelium and air spaces in the soil, it would quickly become waterlogged and less habitable.

The notion of fungal networks is changing how forests are understood. However, when we consider how they encompass other organisms, particularly by forming alliances with plants, they become even more compelling.

Forming mutually beneficial relationships was the secret to life on land. Among the oldest of these ancient unions— arising long before the evolution of vascular plants—are those between fungi and algae, known today as lichens. The lichen alliance broadens the range of conditions that either partner could withstand on its own. Tolerance to extremes of temperature, desiccation, and radiation allows lichens to colonize places uninhabitable to other organisms.

Intimate alliances evolve over eons to bring together different organisms across kingdoms. All work from the same basic principle—combining talents allows for more possibilities than those of an individual existence. Although symbioses were long thought of as an anomaly, or an alternative strategy for survival, the intertwining of organisms is now accepted as the norm. Few organisms, if any, have taken the solo road.

Meet the Mycorrhizas

The survival of most plants relies on their beneficial unions with fungi, or mycorrhizas as they are known. The term mycorrhizas—literally, "fungus roots"—refers to the mutually beneficial relationships formed between fungus mycelia and the roots of most plants. Mycorrhizal symbioses arose early in the evolution of life on land, more than 450 million years ago. Although the first terrestrial plants could photosynthesize, their root systems were not yet sufficiently evolved to extract what they needed from the nutrient-poor primeval soils. Mycorrhizal fungi serve as surrogate root systems. When fungal hyphae attach to plant roots, they increase the roots' absorbable surface area. The fineness and greater length of hyphae give them far more reach than plant roots, accessing pores in soil that roots can't penetrate.

A cache of chemicals makes fungi more adept at unlocking nutrients in soils and organic matter. In return for the nutrients and water supplied by their fungal partners, plants provide fungi with a feed of sugars produced through photosynthesis. It is a good deal, and it works. Mycorrhizal networks are increasingly recognized as orchestrators of plant interactions and nutrient transfer, and are crucial to the growth and survival of most plants.

Although the word "partnership" is often used to describe the mycorrhizal relationship between fungi and plants, the work of a mycorrhizal fungus often goes beyond an individual partner plant. Mycorrhizas can mesh the activities of multiple plants and multiple species. These networks are pervasive, linking up numerous fungus and plant species in sprawling networked webs. Relationships between fungi and plants are likely to be more dynamic than simple give-and-take transactions, encompassing a continuum of flows and exchanges.

It was this sheer pervasiveness of mycorrhizas that, in the late nineteenth century, led German forestry scientist Albert Frank to recognize a vital clue to their function. Although scientists had observed mycorrhizas decades earlier, they had misinterpreted how they worked, assuming fungi were parasites of plants. Given the lowly status of fungi in science and society at that time, the possibility that fungi might be doing something beneficial was not considered. Frank surmised that if their role was purely one of parasitism, then those effects would be evident—the forests would show signs of suffering. That wasn't the case. Frank's discovery was pivotal in renewing ideas about how relationships in forests work.

Mycorrhizas are not only ubiquitous but ancient. Fossils of 400-million-year-old club mosses reveal tree-like coils of hyphae known as arbuscules (from the Latin *arbuscula*, meaning "little tree") within their root cells. Today these relationships are known as arbuscular mycorrhizas or endomycorrhizas (*endo-* meaning "inner") and are so called because arbuscules and other structures are formed inside the cortex cells of plant roots. Arbuscules are the site of carbon and nutrient exchange between fungus and plant. Around 80 percent of plants form these relationships with fungi.

The other dominant type of mycorrhiza is the ectomycorrhiza (*ecto-* meaning "outer"). Ectomycorrhizas form when a fungus wraps a mantle around the feeder rootlets of plants. Hyphae penetrate outwards into the surrounding soil but also inwards between the plant root's cortical cells, providing a large surface area for the exchange of resources. The fungus serves as an interface between the soil and the plant, mediating the intake of materials needed for survival.

Ectomycorrhizal fungi predominantly form relationships with trees. Unlike endomycorrhizal fungi, which mostly produce spores directly from their hyphae, many ectomycorrhizal fungi reconfigure their hyphae into reproductive structures that we recognize as familiar mushrooms. In making a mushroom, a mycelium absorbs water from its surroundings, inflating at a remarkably rapid rate. This explains the sudden appearance of mushrooms after rain. Rainfall and changes in temperature are two main triggers that spur a mycelium to produce mushrooms, but signals from mycorrhizal partners also play a part. While the umbrella-shaped mushroom form is the most familiar, mycelium can also manifest an astonishing variety of other reproductive forms—or "sporing bodies"—such as puffballs and earthstars, corals, jellies, and stinkhorns.

Ectomycorrhizas are extraordinarily successful and have evolved multiple times in evolutionary history. Some ectomycorrhizal fungi are generalists; others are specialists. The famous fairytale mushroom, the fly agaric, grows with various conifers and broad-leaved trees, while the larch bolete is particular to larch. The magnificent Douglas fir trees in the Sol Duc forest are thought to form relationships with more than 2,000 different fungi that provide them with a suite of benefits. Some fungi excel at extracting nitrogen, others phosphorus. Many are adept at capturing water and improving a plant's resilience to drought or heat, salt or toxins. There are fungi that play a defensive role, protecting tree roots from soil pathogens with an arsenal of chemical weapons. A tree that partners with multiple fungi shores up its chances of survival compared to those with fewer companions. In all their configurations, fungal symbioses remind us of the general truth that few, if any of us, can exist alone.

Further north in Canada, in the forests of British Columbia, forest ecologist Suzanne Simard revealed the remarkable extent of fungus-mediated relations and exchanges between trees. She and her colleagues injected Douglas fir trees with radioactive carbon isotopes and tracked how they were transferred between them via ectomycorrhizal networks. Others before her had observed these interactions in the laboratory but not in the forest. Simard's astonishing findings showed that the underground economy of transfer went beyond species or genera: the Douglas firs also exchanged carbon and nutrients with paper birch trees. Simard advanced the notion of tree communities and families —not in a taxonomic sense, but as trees interacting with kin and neighbors. She drew attention to the importance of what she calls "mother trees," that is, older trees. Being old and large, they have extensive root systems and many points of contact, linking multiple species and individuals.

Overlapping networks of multiple mother trees serve as integral hubs of exchange. Their big photosynthetic crowns relay energy to the ground, continually feeding and replenishing the network. Because most mycorrhizal fungi form relationships with multiple tree species, forests are likely to be woven together by diverse mycelial networks. Simard showed how seedlings grow within networks of old trees, plugging into them via their fungal partners and receiving subsidies of carbon, nitrogen, and water. This gives them a head start in the dark depths of the forest floor. It seems that trees are social and benefit by being near kin.

However, not all relationships in the subterrain are as collegial as sometimes presumed. The forest floor is also a fiercely competitive stage, with a continuously changing cast of species and interactions. Fungi live in communities of diverse and complex relationships. Intense competition for space and resources sees some fungi use antagonistic mechanisms to defend territory and obtain food. Some mycologists express concern that popular beliefs about the extent and importance of fungal networks could exceed what is actually known. Metaphor and imagination can help bring science to the people and progress research, but these scientists argue that over-interpreting data, and likening forest networks to human societies could obscure true understanding of how forests function. Few dispute the astonishing discoveries, but some question how the science is interpreted and communicated. Another approach might be to simply appreciate fungi in their own right, in all their wonderful strangeness. It might mean recognizing how they differ from us and accepting that we may never fully comprehend their enigmatic and mysterious lives.

Old and established networks of fungi support the diversity and resilience of the pulsing Sol Duc rainforest.

But long before ecosystems reach this complexity, fungi help create the soils that enable the shift from bare rock to flourishing life. Following a fabulous feast of fungi with Lee that evening, I waved her goodbye and departed the verdant Olympic Peninsula for Europe's least forested country, to see what its fungi are doing.

Rock-Eating Lichens

A daddy longlegs scuttles over my hand and slips into a fissure in the rock. It is one of the first inhabitants to colonize the freshly minted terrain emerging from the retreat of the massive Vatnajökull Ice Cap on Iceland's south-east coast. I'm being blasted by an icy wind and it's hard to imagine a more inhospitable place. As the planet heats up, glaciers are retreating, not just in Iceland but in most regions of the world. Bare rock is exposed in front of glaciers as the ice recedes, the uncloaked terrain offering new habitat for colonization.

Barren rock is one of harshest environments for any life form to eke out an existence. Yet on closer observation, the rock beneath my boots is not as impoverished as it first appears. A palette of lichens is already claiming territory. Map lichens are among the first to colonize, encrusting the rock in sulfur-yellow. Lichens are colloquially known as "extremophiles" because of their ability to withstand extreme conditions and terrains. They grow in the harshest environments imaginable, from blistering deserts to arctic tundra and to salt-sprayed intertidal zones. Lichens are commonly defined as a union between a fungus and an alga. More recently, the discovery that lichens harbor diverse bacteria is shifting definitions as the complexity of the lichen entity becomes apparent. Bacteria are thought to contribute to a lichen's capacity to withstand extremes, and what was once considered a partnership is perhaps more rightly a party.

Although the collective of the lichen entity comprises several organisms representing different kingdoms of life, they are classified within Kingdom Fungi. This is because most of a lichen's mass is made up of fungal cells. The fungus is also unique to each lichen. That is, every lichen has a different fungus that unites with one of a small range of photosynthetic algal partners.

Straddling living and non-living worlds, lichens slowly convert primeval landscapes of rock to soil. The process, called weathering, is both physical and chemical. Minuscule hyphal threads of lichens mechanically disrupt rocks by penetrating tiny fissures. By secreting acids, particularly oxalic acid, lichens gradually etch and corrode rock surfaces. Some lichens fix atmospheric nitrogen and make habitats hospitable for other organisms. When lichens die, their decomposing bodies—along with windblown fragments of organic material—accumulate in the rock's cracks and crevices.

Before plants can get a roothold, the deglaciated rock is colonized not only by lichens but also by invertebrates, and getting there is no easy feat. Spiderlings arrive by aerial ballooning or kiting—an opportunistic riding of air currents on gossamer threads. It seems a precarious and random way to reconnoiter new terrains, but some at least manage to arrive unharmed. Other invertebrates, such as springtails, also disperse by wind or water. They scour the rock, grazing on bacteria, ducking into fissures to dodge the gaze of predatory spiders. Airborne spores that blow in, or are deposited in the guano of birds, colonize any available organic matter. Smidgens of soil gradually accumulate. Deposited spores and seeds wedge in cracks and tap meager nutrients, germinating into the first bryophytes (mosses, hornworts, and liverworts) and grasses.

These elemental landscapes of ice and rock allow us to imagine how the first life forms might have colonized the planet. They also provide an ecological laboratory to understand succession—the process by which the mix of species inhabiting an ecosystem shifts over time.

After hours clambering around on the exposed rock in squalls of icy rain, I feel my nose threaten to snap off, and cut through the wind back to the car. Heading further south, the lava fields are carpeted with the more conspicuous Iceland moss. Despite its common name, it is another lichen. It's known to contain a mild antimicrobial compound and a demulcent (a soothing substance), and is used to remedy inflammation in the mouth and as a treatment for common colds and bronchitis. Among the Iceland moss, other bryophytes, especially gray moss—which really *is* a moss—contribute to the gradual building of the soil layer. Communities of mosses and lichens form thick mantles covering large surfaces and draping over protruding rocks so they appear like large and comfortable cushions. Their short stature keeps them close to the ground, allowing them to withstand ferocious winds that are intolerable to most plants.

The stark and elemental landscapes of glaciers and volcanoes epitomize Iceland. Most vegetation barely reaches ankle height. But that wasn't always the case. Mass deforestation began with the arrival of the Vikings in the second half of the ninth century, with only a handful of relic forests spared. The combination of harsh conditions, frequent volcanic eruptions, vulnerable soils, and intensive sheep grazing prevented forests from regenerating. Forests that once covered somewhere between a quarter to almost half of the country are now reduced to a little over 1 percent. Active reforestation programs are underway to bring back Iceland's forests.

Despite the limited woody vegetation, almost 3,000 species of fungi have been described from Iceland, about a third of which are lichens. As the clouds open and welcome rays of sunshine appear, I leave the rock-transitioning lichens and head off in search of wood-decomposing fungi in the forested 1 percent.

Saprotrophs, Parasites, and Rotting Wood

Bæjarstaðarskógur is a tiny remnant forest of birch and rowan in the Morsárdalur valley to the south of the Vatnajökull Ice Cap. A decent hike over volcanic shingle, black sands, and braided streams takes me into this comparatively aged woodland with its subtly sweet and woody smell of birch. Fenced off from grazing in 1935, the forest now harbors some of the oldest and most established birch trees (150-plus years old) in Iceland. The harsh conditions and limited growing season diminish most of Iceland's trees to shrubs, but those in Bæjarstaðarskógur are somewhat sheltered by Öræfajökull, Iceland's highest volcano. I'm greeted by Eurasian wrens that come to investigate my arrival, along with a common snipe drumming overhead. While various conifer and broad-leaved trees grew in

Iceland in earlier glacial periods, the downy birch is one of only three native trees remaining today. Being tolerant of damp soils and harsh conditions, it grows further north than its broad-leaved cousins.

Downy birches have a special charm. With their often crooked trunks and grayish fissured bark, they're unusually elegant. A carpet of serrated yellow leaves gathers beneath them. Poking through the leaves I spot tiny fibrecaps and a snaking arc of poisonpies. Both are common fungi in arctic habitats. They form mycorrhizas with birches but also with other trees such as willows.

As pioneer plants—those species able to colonize disturbed land—both trees are being used to rehabilitate degraded areas. In the soil-depleted terrains of Iceland, the seedlings' success depends on their colonization by mycorrhizal fungi like poisonpies, fibrecaps, and deceivers. As the succession of species progresses, other fungi such as brittlegills and webcaps plug into the mycorrhizal network. These mycorrhizal fungi are often conspicuous and abundant, and capture attention. However, the fungi that are actually doing the work of wood recycling and soil creation are known as saprotrophs.

Most fungi are saprotrophs. Unlike mycorrhizal fungi that obtain food through their relationships with trees, saprotrophs glean it directly from the wood and other organic matter in which they live. A great swag of digestive enzymes allows saprotrophs to break down molecules such as proteins, carbohydrates, and fats, converting them into simple molecules that they then absorb. This way of feeding by enzyme secretion makes them more like animals than plants: many animals digest food using enzymes inside their bodies, through internal digestion. Fungi also digest food, but they do so outside their bodies, as a form of external digestion.

Wood is incredibly resilient, and breaking it down is a tough task. It's held together by a polymer called lignin that gives it stiffness and structural hardness, and that is difficult to deconstruct. Other polymers called cellulose and hemicellulose, formed from simple sugars, give structure to plant cell walls. Saprotrophic fungi are orchestrators in the continuous processes of decay and renewal, and therefore the process by which soils are formed, but they don't do it alone. Rainwater leaches out the soluble components of wood. Bacteria assist in the decomposition process. Invertebrates mechanically break down organic matter by biting into it and shredding it into smaller pieces. But only fungi can degrade the robust fibers of lignin.

Venturing a little further into the Bæjarstaðarskógur forest, I squat down to examine a fallen birch. I push my finger against its papery bark and it gives. Its softness suggests there are fungi within, actively secreting enzymes and breaking down its woody hardness. A little further along, the trunk is decorated with the semicircular arcs of the ochre bracket fungus, or *gráskeljungur* as it is known to the Icelanders. The ochre bracket's upper surfaces are concentrically zoned in shades of gray and buff, orange and ocher. I roll onto my back and peer at their creamy-colored undersides. They're made up of hundreds of tiny holes or pores, hence the name *polypore* for this type of fungus. Each pore is the opening of a tube in which its spores are housed.

The ochre bracket is just one of the many species of wood-rotting fungi likely to colonize this trunk during the decay process. Mycologists divide the numerous wood-rotting fungi into brown rot and white rot species, based on the type of decay they cause and the appearance of the decayed wood. Each group has different enzymes that break down different woody components. Brown rot fungi often colonize conifers and typically break down hemicellulose and cellulose. The remaining brown residue, composed largely of lignin, becomes part of the humus layer and is an important source of sequestered carbon. Another clue to the presence of brown rot fungi is that the wood cracks and breaks into little cubes that crumble easily into a brown powder.

White rot fungi, like the ochre bracket, commonly colonize broad-leaved trees and decompose cellulose, hemicellulose, and lignin. The resulting residue is white or off-white, often with a fibrous consistency. If you pick up a branch that's been on the ground for a while and a similarly sized one that has recently fallen, you'll notice a difference in weight. Lightness likely indicates the presence of fungi that have dissolved the branch's structural compounds (as well as water loss in the sapwood). Even when we can't see the sporing bodies of fungi, there are always clues to their presence.

Along with mycorrhizal and saprotrophic fungi, a third way that fungi feed—often known as a "trophic mode"—is parasitism. Although parasitic relationships do not benefit both partners, they are another form of symbiosis. Parasites take from their partners—often nutrients or energy—and give nothing in return. Their taking can sometimes kill the host, but most fungal parasites of trees have a vested interest in not destroying their life-support systems. Many are considered weakly parasitic, extracting what they need while ensuring a future supply of nutrients.

Humans have made neat categories for fungus trophic modes but fungi do not always adhere to their assigned roles. Fungi that inhabit wood are categorized as either saprotrophic or parasitic. However, many are more opportunistic and switch between the two modes. The birch polypore, for example, lives mostly as a saprotroph on dead birches but may exploit a weakened living tree as a parasite. Honey fungi (various species in the genus *Armillaria*) do the opposite. They mostly live as tree parasites, but after the death of a tree they can switch to a saprotrophic existence. Having a flexible lifestyle and diet increases the chance of a fungus finding food, and is a clever strategy for survival.

From Wood to Soil

Soils are repositories of life, teeming with organisms. They are the foundations of forests and most terrestrial ecosystems that enable aboveground ecologies to exist. Soil is not complete and ecosystems flounder unless they are brimming with fungi. Plants, of course, do not stand on the surface of the ground. Their root systems penetrate the supportive substrate of the subterrain, from which they draw water and nutrients with the aid of fungi.

By the time a tree falls to the forest floor, it is likely to already have a contingent of fungal inhabitants. Latent fungal propagules—that is, structures that can give rise to a new organism, such as hyphal fragments or spores—exist within a tree's sapwood or bark. They typically have resilient, thick-walled "resting spores" that allow them to bide their time, possibly for many years, until the optimal conditions for germination arise. These "ideal conditions" are thought to relate to water content in the wood. After the tree's death, as the sapwood dries out, the propagules germinate and form mycelia. It's not fully understood how fungal propagules enter sapwood, but it could be via damage from insect activity, bud scars, root wounds or movement through cambium cells.

The way that fungi colonize wood is not always linear but varies according to the sequence of other wood-inhabiting species—lichens, mosses, and liverworts—and the ever-changing chemical and physical composition of the wood. In the initial stages, fungi such as jellies, discs, and pins play an important role, colonizing mainly branches and trunks covered with bark. They are primary saprotrophs that use the simple sugars, starches, and proteins contained in the fresh wood. Through the destruction of the outer bark and wood, these fungi expose the deeper layers.

Wind or animals then deliver the spores of secondary colonizers that intensively decay lignin and cellulose, and gradually outcompete many of the primary colonizers. As the decomposition process advances, the fungus community becomes richer and competition fiercer. With less available space and more demand, some fungi engage in warfare, using their wood-rotting enzymes to mark territory or attack invading fungal colonizers.

The last stages of the wood decay process see colonization by fungi that grow on forest litter. Some of these are known as cord-forming fungi. They amass their hyphae into robust cords, improving the mechanical strength of the mycelium and enabling long-distance transport of water and nutrients. This allows the fungi to form large networks, interconnecting resources and providing long-term nutrient sinks, buffering the forest from nutrient loss. Cord-forming fungi are typically long-lived and traverse large areas, connecting a range of organic matter from twigs to logs.

During the decomposition process, the nature of wood changes. It loses density and changes chemically. The most stubborn components, such as lignin, persist through to the later stages. These changes in turn influence the richness of fungi and other organisms that colonize wood. As a log's wood is dismantled and rearranged, new nutritional niches are created for creatures that feed on dead wood. Hollows and cavities capture pools of water that provide habitats for tiny invertebrates such as rotifers, flagellates, and nematodes. Mosses and liverworts also take advantage of specialist niches that become available during the decay process. In the final decompositional stage other creatures move in, such as mites and earthworms, springtails and myriapods.

A fallen tree can take years or decades to decompose depending on environmental conditions, but in any context there are likely to be dozens, possibly hundreds, maybe even thousands of fungus species that take part in its deconstruction. As conditions change, so does the community of species that inhabits the tree. We won't see them all: many only briefly produce sporing bodies, many are microscopic, and others don't produce any sporing bodies at all. DNA-sequencing technology is revealing the mind-boggling number of species that colonize fallen trees and other habitats. We are only just beginning to discover the intricate details of their lives and interactions, and how they sustain forests.

3

INTO THE
AUSTRALIAN BUSH

Ambling through the south-eastern Australian bush in search of fungi feels light years away from the protocols of the laboratory. Your whole being is called into play. It's a hot fall day when local historian, naturalist, and environmentalist Doug Ralph leads me into the groove of his affinity with Fryers Ridge forest in Dja Dja Wurrung Country in the centre of the state of Victoria. The forest crunches and crackles underfoot, but beyond that the quiet is vast. Eucalypts release their heady cocktail of oils, and the sun is warm on my back.

Doug knows this place intimately. It's his stomping ground. He lives and breathes and feels it. Doug notices details, points out little things. Lichens and liverworts, a solitary sundew, an earthstar. The flat white disc of a huntsman spider's egg case. He tracks and traces, reads history through remnants and remains. Doug finds meaning in the scars and relics of those who were here before us. We cross a trickle of water that is the Columbine Creek, whirligig beetles spinning in circles on the surface. Doug grins at the arrival of a deafening flock of corella parrots, *currup-currup-currup*, and calls back "Well, good morning!" to old friends.

Even in this dry and hungry bush, fungi persist. We drop to the ground to look at tiny mushrooms of the yellow navel lichen staking a claim on a slightly green patch of exposed earth. It's one of the few lichens that produces umbrella-shaped mushrooms, and being chrome yellow, they stand out among the otherwise muted tones of the forest. A little further on we spy a white punk in the fork of a eucalypt. It's a fungus well known to Indigenous peoples in Australia and the Māori in New Zealand. The Dja Dja Wurrung used it as tinder to start fires. For Aboriginal Tasmanians it was food in times of scarcity. They are known as *pūtawa* by Māori, who used it not only to start fires but also to dress wounds, applying it like a protective pad to surround and cushion an injury.

Doug picks up a punk that has fallen from the tree and turns it over. Wasps have drilled into its porous underbelly. "I've seen brown treecreepers pecking the underside of these, but I reckon they were probably after ants inside," he says. Observations like Doug's rarely appear in writing, and I realize how so much local knowledge is held only in the memories of the observers and the stories they tell.

He also introduces me to what he thinks might be the last remaining big yellow box trees in the valley. A handful of these somehow avoided being felled in the 1860s mining boom. Stopping by what he thinks is the oldest, we gaze up into its canopy, then down its smooth trunk. At its base is a troop of small, withered mushrooms known to us as deceivers, their usually reddish-brown caps paler with drying. They're so called as they can change radically in color and form as they age and with exposure to changes in weather, "deceiving" the observer and making it difficult to differentiate them from other small brown mushrooms.

Eucalypts form symbiotic relationships with deceivers and dozens, if not hundreds, of other fungi like webcaps, austral forkgills, and brittlegills. The old yellow box is an important hub of connection between trees in the dry and low-nutrient soils of these forests. Although often regarded as "scrappy regrowth," or "hard to love" in the popular press, these forests matter deeply to Doug. He fights passionately for their protection, sharing his knowledge and founding conservation groups to rally the support of locals.

Doug uses words sparingly, but he speaks as if the whole terrain is listening: the yellow box, the white punks, the deceivers, and the bush. Doug speaks not *about* the yellow box but *to* it, including us all in the conversation. We're distracted by a commotion as a gang of white-winged choughs descends through the branches to investigate our presence. Peering at us through their striking red eyes, they bounce around in a terrain pitted with the shallow alluvial diggings and scars of the goldmining boom over a century and a half earlier. Early images show miners en masse, clambering over and under a landscape almost entirely razed of trees. Every particle of soil was overturned in the mining process, and it's a wonder that any trees or fungi exist today. It suggests the resilience of these forests and their remarkable capacity for regeneration.

This day in the forest with Doug is burnished in my memory. His knowledge of the forest flowed from lived experience. Forests and fungi need those who know them intimately, who observe them over long periods of time and recognize their subtle changes. Less than a year later, aged 66, Doug died of a heart attack. It felt as though a mighty yellow box had fallen. I lamented getting to know him too late. With him went an immense archive of wisdom about the Fryers Ridge forest, and the unwritten stories of a rare being who lived a life in concert with the land.

I wondered who would take his place.

Sensing Fungi

The tools of modern-day science, especially powerful microscopes and DNA technology, radically inform our understanding of fungi, revealing the astonishing diversity of species and their evolutionary histories. However, these technologies mostly approach the understanding of fungi from outside the environments in which they grow. Doug's knowledge of the forest came from reading it through his body, through the senses, in situ. For tens of thousands of years before scientific technologies, human senses were used to locate and understand fungi. Being in the forest with those who are closely tuned to it and have observed it closely over time provides intimate and local understanding of fungi. This knowledge is held in their minds and memories, and passed on through the sharing of experiences and stories to those who are willing to listen.

Observations of fungi don't always begin with the eyes. Smell is often the first sense to awaken when walking into a forest. Older forests are more diverse: they have greater complexity and more processes, and hence smell entirely different from young stands of trees. The ever-shifting mélange of aromas fluctuates with the time of day, the temperature and humidity, and the seasons. Fungus scents emanate not just from aboveground sporing bodies but also from fungal synergies in the subterrain.

Sense of smell is central to the lived experience of the forest. It revives memory and knowledge. Because certain scents are associated with emotions, they can trigger memory more actively than ideas. As a child, my weekend camping trips began with winding down the car window (a time when windows had winders) as we drove into the forest, breathing in the exhilarating scents. Forest scents became synonymous with an overwhelming sense of anticipation of adventure and discovery. I carry that with me today. Opening the window as I drive into a forest, I'm touched not only by the lively headiness of aromas, but the evocative memories they hold.

Being in the forest, and alert to it with the senses, lays down memory in complex ways. If my sense of smell is awake when I discover an unfamiliar fungus, I'm more likely to remember it. The more intense, unusual or complex the aroma, the more strongly it imprints in memory. When I next encounter the scent of that species, I find it easier to retrieve its name.

Although some fungi stink, many of their scents are subtle, and hard to detect and describe. Some are reminiscent of fruits or vegetables such as cucumber, radish, garlic, onion, apricot, artichoke, or bitter almond. Others are foul like rotting potato or boiled cabbage, or they may be fishy, rancid, or fetid. There are those with a chlorine or iodine edge. Many change with age. Sometimes it's hard to differentiate the odor of a fungus from those released by the bacteria breaking it down. Milkcaps are known for their sweep of surprising scents, many of which are not usually associated with mushrooms. I'll never forget my first whiff of a candy cap in the Chuckanut Mountains in northern Washington State, USA. It's strangely sweet, like burnt sugar or maple syrup. Fenugreek milkcaps smell remarkably like fenugreek or curry, yet the curry milkcap, at least to my nose, smells more like camphor. The coconut-scented milkcap, while appearing rather drab, smells enticingly of coconut. Having a food-like odor, however, does not equate with edibility for humans.

My friend and colleague Barbara Thüler is a *Pilzkontrolleurin*, or mushroom inspector. In the Swiss canton of Bern, highly trained mushroom inspectors like Barbara have been saving foragers' lives for over a century by plucking toxic mushrooms from their baskets and sending them home with the edible ones. The first thing Barbara does when identifying a mushroom is closely examine its underside. She then holds it directly to her nose. Then she passes it to me and asks what it smells like. I'll answer "like radish" or "like garlic" or "like chlorine," but sometimes that's insufficient for Barbara, who pauses a moment, raises a questioning eyebrow, then adds "... and?" While many mushrooms have one distinctive odor, others have an amalgam. Her nose is better trained than mine to finely calibrate and distinguish the various olfactory components that coalesce to create its scent. When a fungus does have a detectable scent, it's rare to find agreement among sniffers. Sense of smell, like all senses, is subjective, and even experienced mushroom sniffers can struggle to pinpoint them. As we become less reliant on the sense of smell in the modern world, our olfactory vocabularies have become depleted along with our survival skills.

Sense of touch also reveals reams about the nature and identity of fungi. Touch is ever-present, guiding us through the forest. We can close our eyes, but it's harder to turn off touch. There is no familiar word for lack of touch, as there is for lack of sight or hearing. While we can actively touch fungi to discover their textures, our skin receptors also continuously gauge our surroundings, informing us of subtle changes. Touch tells us not just about textures but about temperature, vibrations, or discomfort—the coolness of the breeze on our faces, or the temperature of the soil, or perhaps the unintentional brushing of a hand against stinging nettles while searching for fungi. Luckily, vibration receptors in my skin once alerted me to a herd of deer charging my way while I was lying in the undergrowth photographing fungi. Touch is deeply primal, providing not just information but emotional connection. The soothing sensation of nestling into moss to photograph a tiny orange mosscap mushroom has, for me, as much to do with the physical and emotional comfort of its softness and coolness as recognizing the fungus or knowing its name.

The subtle textures of some fungi are difficult to describe, but analogies can help. Blewits feel a little like the inner tube of a bicycle tire. A velvet parachute's cap is as soft as its namesake. Brittlegills are so-called for their brittle and breakable consistency, not just of their lamellae (gills) but of the entire mushroom. The blackening brittlegill is especially so, crumbling like Cheshire cheese. Perennial conks (polypores) are often as tough and hard as the trees they grow on. The beefsteak fungus not only looks like a beefsteak but feels like one too. Marshmallow boletes are oddly soft for boletes (fleshy mushrooms with pores on the undersurface) and threaten to melt in your hand like marshmallows on a fire stick. The mushrooms of the porcelain fungus protrude from the trunks of beech trees and glisten with viscid mucus, making them almost impossible to grasp. It's best just to leave them on the tree and be mesmerized by the sunlight filtering through their exquisitely translucent caps, as they drip glop onto your camera lens.

Touching fungi reveals other surprising characters. Even a light touch can cause some fungi to bruise and instantaneously change color. Within seconds some boletes turn rhubarb red or deep cerulean blue, leaving a telltale impression of a thumb or a scratch mark.

Fungi are astonishingly tactile, yet the idea of touching them is alarming to many people, who have misguided fears about toxicity. While there are fungi that are poisonous if ingested, their toxins are not absorbed through the skin by touching them. Some fungi are regarded as repugnant by association with creatures that are supposedly unappealing, sharing slimy or glutinous textures with snails or slugs, or human body fluids. Children are often warned by well-meaning but risk-averse parents not to touch fungi, despite it being instinctive for children to feel and discover their surroundings through the largest organ of their bodies—the skin. Fears of fungi today remain tightly bound to the misconceptions of those before us and can take generations to shed. According to anthropologist Sveta Yamin-Pasternak, there's a fungus known to the Siberian Yupik people called *argaignaq*, meaning "something that makes your hand come off." Despite there being no sign of the fungus living up to its name, touching mushrooms is forbidden by some Siberian Yupik.

Hands-on discovery is a great way for children to learn about fungi. It not only reveals their wonders but deflates unfounded fears that they're poisonous to touch. Once while wandering along the River Clyde in western Scotland with my four-year-old friend Eloise Ferguson, we came across a great snaking arc of mushrooms. Eloise bent down to pat (and talk to) each one. She informed me that the mushroom was called "the prince" and that we could eat it. I was thrilled not only by her knowledge but that she was unafraid to touch the mushrooms. I asked her what they felt like. Eloise stood up, placed her hands firmly on her hips, shot me an exasperated look and blurted, "Well, like princes, of course." I sensed the silent "Stupid!" that she managed to refrain from voicing. Not being privy to that comparative tactile encounter, there was no point arguing with her. I considered asking her what the mushrooms smelled like but decided against it. Should a random prince happen to one day wander my way, I'll expect him to feel slightly fibrous and reek of bitter almonds.

On Country with Yorta Yorta Aunties

Landscapes are read and felt and understood in different ways. Among the people who have been doing it the longest are Australia's Traditional Custodians. I amble among old river red gums and watch the early fall mists rise from Dhungala, as the Murray River is known by the Yorta Yorta people. They are the Traditional Custodians of a 72,000-acre network of eucalypt woodlands and forests, wetlands, and lagoons within the Barmah National Park. Among them are the world's largest river red gum forests and wetlands of international conservation importance. Their lands provide a major source of Yorta Yorta foods, fibers, and medicines. Middens and mounds, culturally marked trees, ceremonial, and burial sites all hold enormous spiritual–cultural significance.

A white 4WD pickup truck pulls up in a cloud of dust and Yorta Yorta woman Sonia Cooper tumbles out of the driver's seat. "Alison! Alison! Hello!" Sonia greets me, wild blond hair falling over her face. "The aunties are here! Come and meet them!" Sonia is the Living Murray facilitator with the Yorta Yorta Nation Aboriginal Corporation who, along with her elders, initiated a project to retrieve and document Yorta Yorta knowledge of fungi across their Country. While there have been various projects in Australia to record Traditional Knowledge of plants and animals, it's the first to focus on fungi. Sonia has invited me along to meet the elders and explore possibilities together as to how the project might unfold.

Sonia's enthusiasm is infectious, and in between stories of the aunties she excitedly tells me about an enormous puffball she has found. We walk down to the river and she introduces me to Yorta Yorta elders Auntie Greta Morgan, Aunty Niecy (Denise) Morgan, Aunty Jan Muir, and Aunty Hilda Stewart. Alongside these elders respectfully called aunties, Yorta Yorta rangers Ralph Hume, Richard Beckhurst, Bonnie Joachim, and Tyler Ferguson work exclusively on Country, looking after Yorta Yorta culture and knowledge.

After a yarn and a sandwich, we wander long the sandy river floodplain where dozens of puffballs, known today as white dyeballs, push up through the bare dry earth. Some are the size of oranges, and the odd one is as large as a grapefruit. Many are distorted and cracked to reveal the mustard-colored powdery spores within. Dyeballs belong to the genus *Pisolithus*, and at least half a dozen species are thought to exist in Australia. We kneel on the ground and cut a firm young one open. Beneath its pale outer layer is a tessellated pattern of tightly packed spore pouches in stunning shades of orange, yellow, tan, and khaki, looking a little like stained glass. The aunties peer closer. It looks beautiful, but as we run our fingers over it, they're quickly covered in a gooey tar-like substance that stains our hands yellow. We all laugh as we try to unstick our fingers.

Dyeballs and various other fungi contain pigments that are useful for dyeing. In Nordic countries, lichens have been used for dyeing wool for centuries. Botanist Carolus Clusius knew of the yellow tincture provided by the chicken of the woods fungus in the sixteenth century. In the 1940s, the French experimented with dyes from mushrooms, but wider interest developed much later, in the 1970s, mainly in Scandinavia, Germany, and North America.

Mycologist Katrina (Katie) Syme and her family lived on a 104-acre property to the west of Denmark, a town in the south of the state of Western Australia, for more than four decades. During their years on the property, Katie explored the fungi among the remnant karri, marri, and jarrah gum trees that grew between granite outcrops. But it was along the northern boundary in a long-unburnt timber reserve that she began observing and recording the fungi in detail, with written descriptions and accompanying watercolor paintings. Katie has also experimented widely with fungal pigments. I ask her whether she has used dyeballs. "Those I collected on our farm yielded shades of brown on wool and golden shades on silk," she tells me. "It is best to collect them well before they mature, when the surface skin wears away and a thick mass of powdery, water-repellent spores develop." The exterior color and appearance of a fungus often conceals its dyeing potential. "Some species of brightly colored gilled fungi contain pigments that produce shades of yellow, orange, and pink, and there's a small leathery tooth fungus that yields green on wool, leaving a lingering odor of fenugreek," Katie explains. "However, just because a fungus is brightly colored doesn't mean it contains dye pigments."

Today, the textile industry is a major contributor to environmental pollution. Growing concern over the toxicity of synthetic dyes for both the consumer and the environment has revived interest in natural alternatives. There are conservation concerns if large quantities of fungi are collected for use as dyes, Katie says, but other natural alternatives are available. "There is plenty of other stuff you can use for making natural dyes, like eucalypt leaves and bark, various flowers, and weeds such as soursob," she points out.

Dyeballs grow throughout Australia and elsewhere in the world, commonly in dry sandy soils. They pop up along track edges and other disturbed areas, often where soils have been compacted, and they can even push through asphalt. Dyeballs form robust mycorrhizas with various trees, including eucalypts and acacias, improving their vigor and survival. I glance at the sluggish Dhungala, and the exposed and floundering roots of river red gums on the undercut riverbank. Extraction of water from the river for irrigation has radically changed the natural cycles and replenishing flows that once delivered nutrients and water to these floodplains. Weirs have slowed the flow of water, obstructed the passage of migratory fish, brought saline groundwaters closer to the surface, and altered wetland ecologies. The combined pressures of water extraction and changes in local climate have diminished the river and its capacity to support the floodplains. River red gums are generally resilient, but these prolonged stresses mean they're faring badly, with three-quarters estimated to be severely stressed, dead or dying. Fungi such as dyeballs are all the more important in these environments where trees struggle to glean what they need from the nutrient-poor soils.

Further along the river we stop by three old river red gums with amazing girth and character. "There's often turtles under these trees," says Auntie Greta. Known as *bayadherra*, the broad-shelled turtle is a totemic protector of the Yorta Yorta people, while the two other local turtle species are a culturally significant source of food. We peer under the trees, but the turtles are somewhere else today. However, we spot diggings in the soil and drop to the ground to see whether they're from turtles, or perhaps wallabies or possums in search of truffles. Auntie Greta gently scrapes the dirt away from the edge of a hole with a stick, unearthing what appears to be a hazelnut-sized stone. It's hard and dry, but as soon as we sniff it we pick up a distinctive fungus odor. We can't be sure if it's a truffle or a dried and withered button stage of mushroom, but it's definitely fungal.

Truffles were known to Indigenous peoples of Australia and various desert nations. These fungi produce their sporing bodies underground, so finding truffles is not easy. Unlike their European counterparts, many Australian truffles appear closer to the soil surface and often protrude slightly. However, it takes keen eyes to spot them. Women are thought to have been the main collectors. They saw the subtle bulges and bumps in the soil pushed up by the truffles as they develop, and used sticks to dig them out of the ground. They were either eaten raw or cooked in the ashes of a fire. Other signs hint at the presence of truffles. In the Pitjantjatjara language of Australia's Western Desert, the word for truffle, *witita*, relates to that for the grey shrikethrush, *wititata*. The bird's call and the appearance of the truffle often occur at the same time, suggesting they respond to the same or simultaneously occurring environmental cues, such as rainfall. By noticing the bird's call, Pitjantjatjara people know to investigate the telltale cracks in the earth that could indicate the maturing of the truffles.

Truffles rely on native mammals to disperse their spores. In these floodplain forests of the Dhungala, swamp wallabies, common ringtail possums, and yellow-footed antechinuses (small mouse-like marsupials) are likely to be important vectors. However, in order to play their part in distributing spores, they must be able to access the truffles in the first place. The habitats of these mammals have been radically reduced and fragmented by the clearing of land for agriculture. Only forested islands remain in a sea of crops and cattle. Consequently, truffles and native mammals can find themselves separated. When mammalian truffle vectors disappear, in time the truffles are likely to be lost as well.

European colonization of Australia saw the wholesale loss of Traditional Knowledge, including knowledge of fungi. Mass genocide, displacement from Country, denial of language and culture, and dislocation of families and communities all prevented the passing down of knowledge. The subsequent introduction of European agriculture resulted in the destruction and fragmentation of landscapes and their ecologies, but I wondered whether the loss of the fungi themselves also contributed to the loss of Traditional Knowledge about them. Shifts in farming practices and especially the extensive clearing of old paddock trees and remnant vegetation using heavily mechanized farming techniques degrades soils for farming. Trees maintain moisture and structure in the soil and reduce erosion. They also form relationships with fungi, including truffles. The loss of trees means the loss of the truffles' mycorrhizal partners, without which the truffles die. The radical shift from native vegetation to crop monocultures could have reduced the availability of truffles for Indigenous people, as well as their capacity to retain knowledge of them.

Over the last two centuries, swathes of Australia's native woodlands, forests, and grasslands have been cleared and modified for agricultural production. Today some 1 billion acres in Australia—just over half of Australia's land—is used by agriculture, but there is a small and slow shift toward less environmentally destructive practices. Progressive farmers are changing techniques to reduce environmental damage: they minimize tilling; retain stubble; reduce fertilizers, pesticides, and irrigation; and rotate stock. Traditional Custodians work alongside scientists, conservation groups, and farmers to restore damaged ecosystems and bring back Country. "Working on Country, you see how sick it is ... it needs to be brought back to life," says Yorta Yorta elder Auntie Greta. "I believe our Country is waiting for us to come back and use our methods of caring for Country. It's waiting for us to heal it."

Sonia's project with the elders involves more than researching fungi on Country. It's also about reviving memory of them. The Yorta Yorta, like many Indigenous communities, talk of storied landscapes and how important they are in connecting to place and to people. Their land is rich with memory and meaning. Knowledge arises from systematic observation of Country and is retained through memory aided by the songs, dances, and stories repeatedly recited and adapted. I ask Auntie Greta if there is a Yorta Yorta name for the white dyeball. "We don't have a name because our ancestors were forbidden from speaking language," she tells me. "They were forbidden from dancing and passing on knowledge about the land in those dances." These are the unspeakable losses that Sonia and the aunties are working to retrieve.

Yet some Aboriginal groups in the central, northern, and western parts of Australia have been able to retain various names for dyeballs. The Kukatja people in the Great Sandy Desert in north-western Australia, for example, know them by several names, including *matjaputi*, *turturtu*, and *puti-puti*, and ate them as food when they were soft and young. The Walmadjari people called them *jurntujarti* and prepared them by removing the outer casing and any powdery spores, then slicing the firm interior and toasting it on coals. It's hard to know whether they were the same species as the white dyeballs we know today—names have changed as taxonomy has become more accurate—but it's likely that several types of dyeballs were used in this way. While different Aboriginal groups used and understood fungi in different ways, sharing of Traditional Knowledge creates possibilities to link together names and revive some of the lost understanding.

The loss of Country and Indigenous cultures, and the changes to the landscape wrought by agriculture, led to the loss of knowledge of fungi. However, there are other reasons why knowledge transfer of fungi differs from that of other organisms. In some of Australia's most remote regions, mycologist Matt Barrett has researched both fungi and Indigenous knowledge of them. Matt describes how fungi, relative to plants, "are the most time-critical group of organisms for knowledge transfer" because the sporing bodies of some species last just a few days or hours, limiting the amount of time that they are available to be collected, and the time that knowledge can be passed on to other members of a group. Fungi are less predictable and sporing bodies are shorter-lived than other food sources, such as plants that produce edible parts over longer periods of time. Hence, fungi are less likely to be a regular staple of diets in the same way as many plants and animals.

Fungus sporing bodies are especially fleeting in Australia's tropical region. "It's a different situation in the tropics compared to temperate Australia," Matt explains. "The combination of high humidity and high temperatures radically reduces the length of time mushrooms persist before they begin to decay, sometimes to just an hour or two." So Indigenous people in Australia's tropical north probably have had fewer opportunities to harvest fungi relative to those living in temperate Australia, meaning that possibilities for transferring knowledge about fungi were likely even more limited. Only fragmentary knowledge of some fungus species might have been passed down.

Māori and Mushrooms

Across the Tasman in New Zealand, Māori Traditional Knowledge of fungi—*Mātauranga Māori o Ngā Harore*—is being rediscovered. As with Indigenous Australian peoples, some records of Māori use of fungi exist in the journals of early European explorers and missionaries. However, much detail has been lost, for similar reasons to those in Australia—oral transfer of knowledge, suppression of traditional languages, and persecution by European colonizers. Fortunately, a team of New Zealand mycologists and linguists is working to retrieve remaining knowledge and revive it among younger generations.

Piopiotahi is the Māori name for Milford Sound in the south-west of New Zealand's South Island. It is the country's wettest inhabited place and one of the wettest in the world. Each year, an average of 250 inches of rain drenches its forests. They're called rainforests for a reason, and if I were a fungus, I reckon it would be a pretty good place to live.

Piopiotahi is contained within Fiordland National Park and is named after the thrush-like piopio bird. The bird has succumbed to extinction, but the ancient forests, with their glades and ravines, harbor 700 plant species, as well as peculiar creatures and diverse fungi. Unlike the plants, the fungi haven't been inventoried, but the rich fungal aromas as I enter the forest suggest they are pervasive. Old logs are dotted with the tiny cyan discs of the green elfcup. Long lobes of speckleberry lichens clothe beech branches. Delicate suede-like tiers of the pagoda fungus poke out from crevices like miniature minarets. Every surface is greened with mosses, lichens, and liverworts. As I clamber through the tangle of epiphytes and vines, a gentle but solid rain sets in. The pattering and trickle of water intersperses with gushes of creeks and rivers, and the thundering of waterfalls. However, within the forest, little rain directly hits the ground. Rather, it is intercepted, filtered through the canopy, and directed down trunks gently to the forest floor. I watch water trickle along a fern frond and drip onto a perfect row of puffballs assembled on a log below. As the droplets of water land on them, each puffs out a cloud of spores in unison.

Pukurau is the name Māori people give to a range of puffballs, some of which were eaten in their young stages before the spores matured and became powdery. Others were used medicinally to relieve pain from burns and to staunch bleeding wounds. Lichens were also used as a soft covering for lacerations. While various fungi were known to be eaten as food, this was probably mostly in times of scarcity, as a supplement rather than a staple food. Among them were *harore*, a species of honey fungus, and *tawaka* or poplar mushroom, which was also used medicinally to treat fever and as an antidote for poisoning by toxic plants. The delicate spine fungus *pekepekekiore* (coral tooth fungus) may also have been eaten. *Hakeke* (rubber ear) is a popular fungus in Asian cuisines, but according to the reports of early missionaries and settlers, Māori weren't partial to eating it. Rather, they recognized its economic worth, and in the second half of the nineteenth century, women collected and exported it in large quantities to China.

Those people lucky enough to have watched a basket fungus emerge from its egg-like immature stage could well have found themselves spellbound. The process by which it unfolds out into the world to release its spores is astonishing the first time you witness it, and no less so on subsequent encounters. This curious fungus forms symmetrical latticed or honeycomb-like structures that look like miniature geodesic domes from a science fiction movie. In its immature egg-like stage, the outer part of the basket fungus was eaten by Māori, but it wasn't actively sought and there are no records of any other uses. Yet Māori endowed this fungus with a remarkable 37 names, among them *matakupenga*, *kōpurawhetū*, *tūtaewhatitiri*, and *whareatua*. Names like *tūtaewhatitiri* refer to its sudden appearance after thunderstorms, while others such as *whareatua* allude to its net-like form.

Humans name organisms to give them significance, especially those that have utilitarian or other values. Naming is satisfying. It provides an identity and a place in the various categories we use to give order to nature. A name provides a reference point to talk about the same species. Yet many species are known and named without having a particular use. The basket fungus is unusual and visually conspicuous, and it has a fetid odor when mature, so it's likely to be noticed. Still, while there are many distinct Māori dialects across the country, 37 names seems extreme for a fungus that had little utilitarian value. Perhaps this tells us that before the introduction of cattle, it was more abundant and more frequently encountered and talked about. In this way, Traditional Knowledge not only reveals the names and uses of particular fungi, but it hints at how environments and their fungi might have changed over time.

The categorizing of organisms by Māori and other indigenous cultures often works on different logics from the hierarchical Linnean classification systems of Western science. While the two systems are independent, each provides scientific and cultural contexts and can complement the other. Indigenous naming is more likely to stem from pragmatism and practical use than from evolutionary relationships. It's usually more flexible and open to interpretation, less hierarchical, and often strongly contextualized in place. Sometimes a single name refers to multiple species, suggesting the uses of these fungi might have mattered more than their naming and categorization, or differences between species might simply have been too subtle to have been noticed. Mycologist Peter Buchanan describes how in Māori Traditional Knowledge, fungi were not recognized as a distinct or unified group and there was no generic term to refer to them. A collective Māori term for fungi was only recently selected. "Choosing an appropriate Māori name for fungi was not straightforward and involved a lot of discussion with specialist translators," Peter explains. "Initially *harore* was advised, but later this was considered to be too restricted to mushrooms, and instead *hekaheka* was adopted as a broader term that earlier mainly referred to molds."

When cultures are disrupted, and traditions and memory are lost, knowledge held in oral stories can disappear forever. Peter has been instrumental in reviving Māori Traditional Knowledge of fungi and making it accessible to the broader public. He initiated a project to collate Māori knowledge, working alongside Māori educationalist Georgina Stewart and Māori language translation specialist Hēni Jacob. The *Ngā Hekaheka o Aotearoa* (Fungi of New Zealand) project involved both direct interaction with Māori elders and younger Māori, as well as sifting through historical records of Māori use of fungi. Although most knowledge appears to already be lost, naming fungi in Māori language can change the way nature is experienced and understood.

When language is lost, so are the stories, insights, and traditions accumulated over hundreds or thousands of years. Language is one key to unlocking historical knowledge of fungi. Peter and his team, and Sonia and the Yorta Yorta elders, are gradually reviving it, one fungus at a time.

4

NO SUCH THING
AS A BAD FUNGUS

Clément Léandre arrived an hour early at my fungus workshop in Wagga Wagga, in the Riverina region of New South Wales, south-eastern Australia. I was racing against the clock to be set up and ready on schedule when I heard his tentative knock. I had been hoping for a little more time to prepare, but I opened the door. It turned out that Clément had driven five hours north from Melbourne, departing at 3 a.m., apparently with a very pressing question about fungi.

Clément was studying horticulture in Melbourne. He asked if he could help me set up, so I put a handful of labels with fungus names in his hand to assign to the specimens on the table. Clément grinned self-consciously and took the labels. In a barely audible voice he told me about his studies, moving swiftly around the table, examining and labeling each fungus. Clément was fascinated by fungi and the interactions between them and plants. He'd come to the workshop because he questioned the approach of his horticulture lecturer and was keen to hear other perspectives.

"You know, the main thing I've learned about fungi is how to kill them!" Clément sighed. He told me that the only fungi his course addressed were those considered pathogenic. "With all the players in the ecosystem and all their interactions, is it not fantasy to isolate a single species as the cause of a problem?" he asked, shaking his head. He wanted to know why fungi are not regarded in their own right and are studied mostly through the limited lens of plant pathology, as a threat to be managed or destroyed.

The history of how mycology developed in Australia provides a clue as to why. The scientific study of fungi has always been tightly linked with agriculture in Australia. Pathogenic microfungi, such as rusts and smuts that infect crop monocultures, especially cereal grains, were—and still are—a prime focus of Australian mycological research. Systematic mycology in Australia was formalized in 1890 with the appointment of Scottish agricultural scientist Daniel McAlpine to the Department of Agriculture in Victoria. He was specifically employed to research fungal rusts and other plant pathogens. By the end of the nineteenth century, departments of agriculture and agricultural colleges were established in all Australia's colonies. In 1926, after 25 years of nationhood, the Australian Government's scientific agency, the Council for Scientific and Industrial Research (now the Commonwealth Scientific and Industrial Research Organisation or CSIRO) was founded with support from Britain's Empire Marketing Board. The expansion of industrial agriculture since the 1950s has seen mycological research focus on microfungi associated with plant and animal disease; pest control; food spoilage; forestry; and, since the 1980s, biosecurity.

Early government science agendas in Australia were based on concerns for "the national interest." Agricultural and forestry sciences were economically driven; and, as historian of science Libby Robin notes, the CSIRO's funding "depended on convincing politicians of science's role as a handmaiden to economic growth." Microfungi had economic significance in the agricultural industry, but the larger macrofungi—those that produce the familiar mushrooms and other sporing bodies—are less commonly found in agricultural systems. We now know that macrofungi are crucial for the functioning of ecosystems, but this was little known early on.

While macrofungi have been collected for science in Australia since British colonization, it was initially in the context of the amateur naturalist, not the professional scientist. Field naturalists were the primary observers, recorders, and collectors of fungi. This work can only be done in the field, in nature, not in the laboratory. Although field naturalists are often associated with being "amateur" this refers to them being unpaid, not to a lack of knowledge. Their specimens were sent to mycologists in England and elsewhere in Europe for identification, as there was no recognized taxonomic expertise in Australia. Most of the early type specimens (the original specimens used to describe a species) remain in the fungarium at the Royal Botanic Gardens in Kew. Today, scientific understanding of fungi begins with giving them names and categories, and discovering their evolutionary histories and relationships. This has broadened in recent decades to include areas of research such as fungal ecology, biogeography, and conservation of macrofungi, but few paid positions exist.

Back in Wagga Wagga, Clément returned from the fungus display table within minutes, having assigned all the labels, and asked for more. I glanced at the specimens and was astonished at how quickly and accurately he'd labeled them all. Clément was not only good at fungus identification; he was also an expansive thinker. We discussed the tangled ecologies of fungi and their environments, and the challenges of balancing the overlays of human needs and desires. He sought to understand the complexity of interactions, rather than simply seeking the "outcomes" expected by land managers, and recognized that relationships between species are intricate and nuanced. As part of a younger demographic of the myco-curious, Clément was one of those rethinking some of the assumptions around how we understand fungi and manage ecosystems.

His concern echoed that of an arborist I'd met a week earlier. The arborist had been contracted by a local council to manage its public parks and gardens. He lamented how a fundamental misunderstanding of fungi combined with an over-inflated sense of public risk were proving catastrophic for the council's ill-fated trees, some of which were listed on heritage registers. Given everything we now understand about fungi and their importance in holding ecosystems together, I wondered how this could still be.

The arborist considered the council manager's restricted thinking to be the biggest threat to the survival of its trees. "He would ask 'What's *wrong* with the lawn?' if mushrooms popped up," said the arborist. "And when clumps of sulphur tufts grow at the bases of trees, he'll say to me, 'I don't like the look of those. Watch that tree—I reckon it'll have to go.'" This fungus is a saprotroph that breaks down organic matter, releasing nutrients. If anything, it provides benefit, not harm, to the tree.

The arborist and I had a long conversation about the thousand ways a fungus, or a tree, or a park, or a forest can be viewed. Every view is only ever a partial one. We rely on our perception and understanding, along with ethics and intentions, to make judgments. We make an effort to be objective, but our views inevitably sit within our ideologies. Given fungi have historically been maligned, there's a lot of undoing required to shift the frames of some people's thinking so that fungi can be recognized as beneficial organisms.

Land managers and councils also have the difficult task of juggling the needs and whims of numerous "stakeholders." The ecology and long-term viability of a park or garden is usually overshadowed by concerns for safety or aesthetics.

"Solution-driven efficiency" can override complexities and subtleties that don't fit within the limitations of short timeframes, budgets, and reporting schedules. When dealing with a scapegoat such as a fungus, reinforcing it as the enemy is often the easiest and quickest path to an outcome. It usually sees the unnecessary destruction of the tree and the fungus as the most economical option.

After the Wagga Wagga workshop, I was packing up the specimen table when I heard a tentative knock at the door. The knock was familiar. I called out to Clément to come in. It was already growing dark and I knew he had a long drive back to Melbourne, but Clément had further questions. He asked if he could help me pack up, so I suggested he collect the labels with species names from the specimens on the table. Clément grinned a little less self-consciously and started to collect them. He then paused and asked, "What if instead of each specimen having a label with a name, it had a label describing what it did? They could read, for example, 'recycler,' 'drought proofer,' 'soil stabilizer,' 'fire starter,' 'nematode strangler,' 'bird-hollow creator,' 'medicine source,' 'food provider,' 'post-fire soil restorer' ..." It was a brilliant idea and a clever way to increase appreciation of what fungi do. I watched his smile slowly broaden.

Clément's suggestion reminded me how different languages take different approaches to knowing and naming organisms. I recalled anthropologist Richard "Nels" Nelson telling me how Alaskan Koyukon people name birds based on a characteristic behavior. What an organism *did* is more significant to the Koyukon than a noun-based name that described its appearance. In doing so, they capture an organism's dynamism and interactions with their worlds. As with many indigenous cultures, naming, and categorizing organisms relies on different frameworks from those of Western science, often with a more immediate and localized pragmatism. Fungi do a lot of things in ecosystems, mostly out of sight and beyond our awareness. Clément was onto something.

Reframing the Scapegoat

Histories of misunderstanding about fungi stretch back centuries, and the early association of fungi as agents of disease continues to shape thinking about fungi today. In Australia, fungi continue to be seen as an enemy of agriculture, yet modern industrialized agriculture creates and perpetuates "fungal diseases" by planting monocultures. Unlike polycultures, which incorporate multiple species and attempt to mimic the diversity of natural ecosystems, monoculture cropping is the practice of growing one crop. Monocultures damage soils and destroy fungi; consequently, crops require more fertilizers and irrigation, and are more susceptible to disease.

This is not to understate the damage microfungi can wreak when agriculture or forestry are mismanaged. Rather, it is to question the pathologizing framework that defaults to the position of fungi being problematic, disease-causing, and in need of control or elimination. Most fungi are not pathogenic. Those that are rarely reach problematic proportions in natural ecosystems, and most "fungal outbreaks" are a symptom of poor human management.

I wonder what it is about fungi that people find so offensive. What could be so alien about these organisms with whom we share a not so distant evolutionary past? Or is it precisely that paradox that grates, the sneaking suspicion that they are not as distant as we might suppose? Fears around fungi persist. Even today, gardening magazines are funded in part by advertisements for antifungals. Arsenals of fungicides and other pesticides line the shelves of nurseries. Although less aggressive and friendlier approaches to gardening have sprouted in recent decades, the chemical industry retains its grip. Fungi persist as prime suspects for perishing plants and inexplicable witherings.

I'm regularly asked how to eradicate mushrooms from the lawn, the tree belt, the compost pile, the pot plants, the sandbox, the flowerbed, the driveway, and the vegetable garden. There are constant concerns about the risk fungi pose to pets and children. Others are affronted by lichen on the gate, the roof tiles, the fence, the path, the garage door, and the mailbox. The list goes on. The default response to something perceived as a threat is to eliminate it. On a more alarming scale, industrialized agriculture is firmly ensnared in a cat-and-mouse game: as the use of antifungals steadily increases, so do pathogenic fungi that are resistant to fungicides. Despite decades of class actions and lawsuits concerning the destructive effects of its products on biodiversity, ecosystems, and human bodies, the agrochemical industry remains deeply invested in ecocide.

Another challenge to shifting the focus on pathogenic fungi is the lack of language with which to talk about them or, rather, to talk about their beneficial nature. Centuries of negative associations in the English language and a scientific focus on their destructive capacity have fueled a vocabulary of disdain. Few poets or creative writers have celebrated their beauty or benefit. Writers and poets—mostly from the nineteenth century—such as Arthur Conan Doyle, John Keats, Alfred Tennyson, and Emily Dickinson—regarded fungi as symbols of degeneration and death. Among the better known is Arthur Conan Doyle's fraught association between fungi and disease in his novel *Sir Nigel*: "The fields were spotted with monstrous fungi of a size and colour never matched before—scarlet and mauve and liver and black. It was as though the sick earth had burst into foul pustules; mildew and lichen mottled the walls, and with that filthy crop, Death sprang also from the water-soaked earth."

Fortunately, some people revere fungi. One of the most enthralling aspects of fungi is the sudden appearance of mushrooms. The wonder and power of this phenomenon is captured in the word *puhpowee* from the Native American Potawatomi language. It translates to "swell up in stature suddenly and silently from an unseen source of power." According to North American Potawatomi Nation citizen and biologist Robin Wall Kimmerer, it can refer to "the force which causes mushrooms to push up from the earth overnight." Concepts for which there are no words in one's own language are beguiling. *Puhpowee* is capturing imaginations in the myco-community not only because it is powerful, but also because it represents more than mushrooms. It inspires ideas about flourishing and vibrancy, emergence and transformation, and counters limited notions of death and disease.

Language develops for the things we notice or that are significant to us. Specific and specialized language is necessary to describe them. So the lack of vernacular terms for fungi in the English language means these organisms lack not only our awareness but our regard. As the world becomes more mediated by technology, and sensorial encounters with nature dwindle, we risk losing language for things that are important or that inspire our imaginations.

Unexpected Vectors

On the other side of the world from Australia, a tiny forest grows a stone's throw from the old castle in the settlement of Nidau in north-west Switzerland. Just a few hundred square yards in area, it is known simply as the Erlenwald or Alder Forest. Lake Biel borders its north-western side. Houses and tennis courts butt up against its others. Walkers and swimmers wander its paths en route to the lake. Despite the forest's small size, it is home to a handful of magnificent alders, beech, willow, and ash. The only green oasis in the town, it is just large enough to allow you to take a deep breath, stare up into the canopy, and imagine you are somewhere deep in a remote forest. If you're quiet enough, you might spot a woodpecker or squirrel, or if you keep an eye on the ground you could spy a troop of shaggy inkcaps standing tall like forest sentries.

Late one fall, as the trees relinquished the last of their last leaves, I wandered through the tiny forest only to be assaulted by the blast of leaf blowers. A team of men in high-vis gear aimed their ear-splitting instruments at anything vaguely biological. It seemed that even within a forest, tidiness was the priority. Stripping every last leaf from the forest's paths, they revealed clumps of honey fungi (*Armillaria mellea*) sporing bodies clinging for their lives to the bases of tree trunks and stumps. While honey fungi are pretty good clingers, their spores are not. That's the whole point, after all, if you rely on wind for spore dispersal. When you stand an inch or two from the ground, only the tiniest currents of air aid the process. You're lucky if your spores disperse more than a yard or so. A sudden 150-mile-per-hour squall courtesy of a leaf blower, however, could be a helpful catalyst.

When a tree becomes unwell, it's easy to point the finger at a fungus. However, this overlooks the complexities of forests and the role of humans in the environment. Honey fungi, for example, are often scapegoated as a cause of tree death, but their flourishing is often a symptom of complex changes in their habitats. When humans homogenize environments by prioritizing aesthetics over ecology, the complex entanglements that exist between species can be disrupted. Honey fungi rarely kill healthy trees growing in optimal environmental conditions. Rather, the honey fungus—and others such as the red-banded polypore, the tinder polypore, and the popular edible oyster mushroom—grow on weakened trees. Like predators that seek out feeble prey, some fungal parasites are thought to weed out the old and decrepit among a stand of trees.

Stresses such as soil compaction and contamination from poorly practiced forestry, drought and high temperatures, and air pollution can all sap a tree's vigor. Trees are resilient and resistant, but when continually exposed to multiple stresses they gradually wane, making them an easy target of parasitic fungi. Some fungi, such as the chicken of the woods, can infect healthy trees. But most fungi, including the honey fungus, tackle trees that are already in decline, either as a natural part of the aging process or from environmental or human stresses.

The honey fungus has several secrets to survival. One of them is the ability to felt its mycelium into long ropy strands known as rhizomorphs. Heftier than individual hyphae, they transport water and nutrients over longer distances. Adapting your mycelium into these more robust rhizomorphs is a clever way to cover more terrain, claim more territory and infect more trees. Its spores are possibly just a backup strategy or insurance policy should its mycelium meet its fate. Nevertheless, I wondered how leaf-blowing humans could affect the spore dispersal of fungi, especially those considered "problematic."

I took a closer look at the clumps of honey fungi, many in great gangs of dozens of mushrooms. They grow in caespitose clusters, which means their stipes (stems) sprout from a common base. As such, the mushrooms crowd together with overlapping caps. Consequently, the lower mushrooms are usually dusted with the white spores of those above, appearing as though fairies have given them a gentle powdering with icing sugar. The mushrooms in these clusters in the Erlenwald were mature enough to release spores, yet all were unpowdered, some with their caps blown clean off. I glanced around, making a low 360-degree swooping survey of the undergrowth. No fairies. I then made a shortcut dash through the forest to overtake the leaf-blowing brigade and inspect some unblown fungus clumps. Sure enough, these were beautifully dusted in white spores.

It doesn't take a genius to work out that a high-velocity gale emitted from a leaf blower could dislodge spores. But where and how far do spores go? Or do spores not matter at all if mycelium determines survival? And what of those spores stuck to the men's orange jumpsuits or the linings of their lungs? Were they the real vectors that inadvertently relayed spores to the next forest or garden?

Spores get blasted by blowers, but it's not much fun if you are a grasshopper, beetle, or frog either. How has the Leaf Blower Revolution changed the lives of those creatures less offended by a fallen leaf than humans? Fungi don't have ears, but I wonder how the roar of two-stroke engines and the accompanying spewing of spent fuel, of carcinogens like benzene, might affect finely tuned networks of fungi. Such things are measurable by the tools of science, but they need to be noticed in the first place. I suspect no one is surveying the honey fungus in this forest. What happens if, over time, the fungus colonizes every tree? What happens when every leaf, plus the many species of fungi that keep the honey fungus in check, gets blown into oblivion? Fungal diversity and consequent competition for resources is what keeps any one species from dominating and gaining the upper hand, but all the other fungi also need places to live. The greater the availability of specialist niches and habitats provided by a diversity of organic matter, the more species of fungi working to shore up the system.

The advance of the honey fungus through the forest is most apparent to us in the short period when it produces mushrooms. However, beneath the soil and within wood, the rhizomorphs of the honey fungus continuously extend their long-lived connections between trees and islands of woody matter, expanding the fungus's territory. Will the trees gradually succumb? Will the workers eventually swap leaf blowers with chainsaws and fell the elders? What will they do when there are no more leaves to blow? Perhaps they will gladly take up tennis on a synthetically surfaced court.

I'm not aware of any testing of leaf blowers or high-vis vectors for spore dispersal but, given the global ubiquity and burgeoning population of leaf blowers, the consequences of these machines on spore dispersal could be worth thinking about. Perhaps then, with the trees and fungi just a distant memory, rakes might seem like a good idea.

How might we respond to these situations? Fungi rarely figure in the equations and models of how ecosystems are managed. What happens over time when we manage only for those species that are visible to us or perceived as important or appealing? In the Erlenwald, fungi are both victim and enemy of the tidy. But back in Australia, a newly discovered fungus has blown up into a more deadly threat.

The Poison Fire Coral Saga

Australians are often enamored of the villainous and the venomous, or at least with talking about them. Nothing beats a good story, especially with a killer as the protagonist. The flame fungus—or, as it's more commonly and dramatically known, the poison fire coral mushroom—appears to be the latest menace to threaten Australia's inhabitants and unwary visitors.

In 2019, fungus enthusiast Ray Palmer was exploring a less accessible part of Cheepi Creek in suburban Cairns, in Australia's tropical north. Between the roots of an old quandong tree he noticed an unusual red fungus. Ray was curious. He has a pretty good grasp of local fungi but he'd never encountered this one, although he'd seen something similar in books and so had a reasonable clue as to its identity. To be certain, he sent it off to mycologist Matt Barrett at

James Cook University for confirmation. Sure enough, Ray had made the first reported discovery of the poison fire coral mushroom Down Under.

First discovered and described (originally as *Hypocrea cornu-damae*) in 1895, the poison fire coral mushroom is thought to originate in Tibet. It also grows on the Korean peninsula, in Japan and Java (Indonesia), and more recently has been found in Thailand and Papua New Guinea. Whether this fungus arrived in the Australian tropics from elsewhere is not certain. It could have been around all along and, despite being conspicuous, had simply been overlooked. Since Ray's discovery it has popped up further north in Mossman Gorge and the Daintree Rainforest, and is likely to appear elsewhere. As its name suggests, the poison fire coral mushroom emerges from the subterrain as crimson-red "flames" or "fingers." It is distinctive and extraordinarily beautiful.

The poison fire coral mushroom has all the attributes of a good story. It is novel, flamboyant, and, most importantly, deadly. Reports of the poison fire coral mushroom spread like wildfire in Australia and beyond. Within days headlines appeared, each one more lurid than those preceding: "Terrifyingly lethal poison fire coral fungus"; "This deadly mushroom physically shrinks your brain when eaten and will poison you if you even touch it"; "This deadly mushroom can literally shrink your brain—and it's probably more widespread than we thought: Australia adds one more horrifying species to its roster." Across continents, alarmist tales popped up in European newspapers and in the newsletters of North American mycological societies. As the token Australian in a foreign land at the time, I was inundated with cautionary warnings—some tongue-in-cheek, others not. At first, I was excited that Australia and fungi had made the news, but I quickly realized that alarmist hype does fungi few favors.

Sensationalist stories aside, let's consider the facts. The poison fire coral mushroom is commonly reported as being poisonous to touch, yet no evidence exists to support this claim. While peeling of the epidermis (outer layer of the skin) is a reported symptom, this is due to poisoning from consuming the fungus, not from touching it. It seems the presumption of contact toxicity is a misinterpretation by an overzealous journalist that has since degenerated into a fanciful and widespread myth.

The poison fire coral mushroom does contain a lethal mycotoxin. Fatalities are reported from Japan and South Korea. People have died from eating minute quantities. Accidental poisoning is probably a result of it being confused with some edible species of the coral fungus genus *Clavulinopsis* that are consumed in Japan, or *Cordyceps sobolifera* used in traditional Chinese medicine. It is hard to find accurate information on the number of fatalities, but at the time of writing, reputable reports suggest fewer than ten people have died. One report told of thirteen people ingesting it in Japan, resulting in two fatalities; another report said that five people drinking sake made with this fungus resulted in one death; in South Korea one person died from the four who drank tea made from it. In the reports from Japan and South Korea, most people who consumed the fungus did not die.

While the effects of touching this fungus are questionable, the possible consequences of eating it are not. Acute symptoms can appear within 30 minutes of consuming it and can include dehydration, abdominal pain, diarrhea, and severe vomiting. The toxins, known as trichothecenes, can also cause the inhibition of protein synthesis (a vital life process); destruction of bone marrow; and a suite of other symptoms including fever, low blood pressure, changes in perception, disturbance of consciousness, and progressive severe respiratory distress from lung necrosis. Symptoms can progress to include hair loss; peeling skin on the palms, soles, chest, and face; and brain atrophy. Death can result from multiple organ failure, including liver and kidney failure, and blood clotting. In short, it is a horrendous way to die.

Without trivializing the tragedy of those who met their fate from the poison fire coral mushroom, keep in mind that fatalities have also resulted from consuming cow de-wormer, strawberries, and marshmallows. And paracetamol—a lot of them. Australians poison themselves with far less imaginative substances than the poison fire coral mushroom, with paracetamol being the favorite. Government-approved pharmaceuticals (especially psychotropic medications and analgesics) constitute the majority of calls regarding suspected poisonings to Australian poisons information centers. Toilet cleaner, personal care products (hand sanitizer, cosmetics), and benzodiazepines (sedatives) are also common.

A quick scan of the poison information centers' annual reports clearly reveals that more dangers lurk within our homes than beyond. Exposure to things from nature such as insects, plants, and mushrooms collectively constitute a small minority of calls to these centers. While plants attract much less attention in the press than fungi for their poisoning potential, they typically result in many more calls to centers than mushrooms. Moreover, the 2019 report from the poisons center in the north-eastern state of Queensland records insecticides as causing almost three times as many calls as the insects themselves. Further south, in Victoria in 2018, there were 226 calls for suspected mushroom poisoning, fewer than those for soap, glow necklaces, or nappy rash products. The gruesome focus on fungi is wildly disproportionate.

If people eat the poison fire coral mushroom they can become sick or die, but the mushroom is not the colorful villain that the press would have us believe. The reasons why some fungi evolve such powerful toxins is not fully understood, but what is known is that it's not to poison *Homo sapiens*. What fungi do—including the poison fire coral mushroom—is give life. They contribute to the functioning of ecosystems and our survival. They provide architecture in soils, drive energy cycles, and recycle nutrients and release them, making them available to other organisms. They filter water. They provide habitat and food for myriad organisms. In our parks and gardens, fungi protect plants from disease and improve their resistance to drought. They delight and inspire. The list goes on, as described throughout this book. They have better things to do than lure *Homo sapiens* to their deaths.

The poison fire coral palaver has little to do with the dangers associated with this fungus and a lot to do with entertainment and inciting controversy. It might all seem like a bit of a thrill in the spirit of good fun, to quicken the pulse, but nothing to get hot under the collar about. The problem is it generates fears and fantasies that are accepted as truths. Overly vigilant councils already unnecessarily eradicate fungi from public parks and gardens. Yet they seldom remove the rhododendrons, hydrangeas or oleander, or the children's seesaw. Deliberately provocative and misleading reporting of a life form that peacefully exists perpetuates public misunderstanding, erodes conservation, and builds barriers to nature.

Perhaps it's time to take a deep breath, calm down, and inject a little reality into the situation. Imagine what could be gained if we considered an inspired coexistence with this stunningly beautiful organism. Imagine if we took the time to meet and get to know the fungus, rather than read the horror stories about it. The poison fire coral mushroom has captured our attention. That's a good thing. It's not asking people to invite it into their lounge rooms, but simply to acknowledge its existence and let it be.

5

FUNGI, FIRE,
AND ICE

I wasn't prepared for my 33rd trip to Gariwerd, in Victoria, Australia. The landscape here in the state's west was so utterly changed it was barely recognizable. I'd been visiting Gariwerd, also known as the Grampians National Park, since childhood but all my familiar landmarks had vanished. The messmate with its lightning scar. The dell of pyramidal cherry ballarts. The old hollow eucalypt full of roosting bats. With the vegetation razed, strange new rocky projections jutted from the ravaged terrain. Disorientation crept in. I bent down and scooped up a handful of blackened dirt. It was strewn with tiny charred and crumpled bodies. Broken exoskeletons. Nothing moved. The landscape smelled and felt like death.

A lightning strike ignited a bushfire at Warrinaburb (Mount Lubra) on January 19, 2006. An intense inferno soon punched up into the stratosphere, generating its own weather system, melting metal, and filling lungs with choking smoke. When it was extinguished a month later, 455,000 acres had burnt, including nearly half of Gariwerd. But it was relatively small in the growing trend of large and highly destructive fires fueled by the vagaries of a changing climate. The Black Summer fires of 2019–20 burnt more than 17 million acres—an area larger than the United Kingdom—or about 40 percent of Australia's eucalypt forests. These unprecedented fires operate at another scale, intensity, and unpredictability. Summer, according to the "authorities," no longer exists. It is now the "fire season." As is often the case after major fires, tolls of human lives and property were tallied. Devastated ecologies were not.

I knew these sandstone mountain ranges well—until the fire. I'd spent the previous three years exploring and photographing them for a book about Gariwerd. Their varied habitats harbor a huge diversity of species. Many are found nowhere else. Now the silence was surreal. Fire brings such radical change. The confrontation of blackened landscapes can make it difficult to reconcile the sense of loss and grief. I was tempted to close my eyes and hold it in my imagination as it once was. How would I be able to reconnect to a place where everything that was once reassuringly familiar, yet endlessly surprising, was gone? Memories collided. The ancient red gum tree that I'd camped by since early childhood had perished. On one trip, aged seven, I'd filled my tent with the giant luminous mushrooms of ghost fungi. I was spellbound by their otherworldly glow and woke the next day with them squashed through my hair. Other times I'd force myself to stay awake, entranced by the night sounds and the impossibly starry sky. Now a stubble of blackened trunks extended across the range as far as I could see. All felt hopeless. Until I dropped my gaze.

On the ground was a pinky-orange fuzz. Peering more closely, I saw the lumpy crust of the cup fungus *Pyronema omphalodes*. It was the only sign of life. Expanding my focus, I saw that it was widening its territory, bracing the flimsy surface with a protective integument. I puzzled how this fungus could possibly have survived the inferno. How could it eke out an existence in the inert and fickle soils, now reduced to little more than ash? It was only days since the fire had been extinguished, yet I knew that with the first rains, tiny shoots of hopeful green would appear among the blackness. In time, surviving trees would sprout cloaks of new growth. But before the green returned, the pinky-orange fuzz of a fungus was kickstarting the long, slow process of recovery.

Pyrophiles

Australia is one of the most fire-prone countries in the world and fire is a natural phenomenon of the Australian bush. For thousands of years before European settlement, Aboriginal Australians used fire to manage certain types of grasslands and woodlands, regenerating grasses to attract herbivores for hunting. Many fungi, animals, and plants evolve to cope with fire. Those that live in systems with predictable fire regimes are adapted to survive or rapidly recolonize following fire.

Fungi that appear in response to fire are described as pyrophilous (fire loving), carbonicolous (charcoal inhabiting) or phoenicoid (growing from ash, like the phoenix, the mythological bird that rose from the ashes). They can tolerate the extreme temperatures, and the physical and chemical changes that fire brings. Some actively need fire as a stimulus for renewal. Others appear after fire but don't require it to trigger reproduction. Although scientists have recorded the appearance of fungi after fire for over a century, Aboriginal Australians are likely to have observed them for thousands of years. How these fungi manage to colonize and subsist in burnt habitats is only slowly becoming understood.

Fungi are often the first life forms to appear after fire. While some, like the leathery sawgill, produce large conspicuous mushrooms, many pyrophilous pioneers are tiny disc fungi that cluster together, closely hugging the ground. With most fungus competitors having perished, pyrophilous fungi take advantage of the bare terrain and the pulse of nutrients released by the fire to spread their spores. These fungi have a range of fire-proofing strategies. Many have hardy spores that lie dormant until triggered by heat or rain. Others respond to the increase in soil alkalinity following fire. Fire sterilizes soils, killing other microorganisms, so opportunistic fungi can capitalize on the reduced competition and claim new territory.

As the collective of early colonizers grows, the architecture of the friable post-fire soils is slowly rebuilt. Fungi add structure and stability through their mesh of mycelia, aggregating soil particles and holding the earth together. Speed is important when colonizing new territory, as are numbers. Their window of opportunity is short and they often appear en masse, painting the blackened forest floor with hues of orange and pink, purple and brown.

Gradually the forest floor transforms. The physical presence of pyrophilous fungi creates microhabitats that shelter seeds and other post-fire colonizers such as liverworts and mosses. Some fungi concentrate nutrients that are tapped by these photosynthetic pioneers. As the organic layer accumulates, it is in turn recycled by fungi. Decaying sporing bodies themselves contribute to the mix. Nutrients are released. After rain, water flowing over the surface is slowed and gently infiltrates the soil. Water retained in the system enables invertebrates and other animals to return as the forest begins to recover. Habitat complexity increases and new fungi colonize. The newcomers' strategies for survival differ from those of the opportunistic colonizers. Some are more cooperative, and the reciprocal relationships they form with recovering plants promote stability as they slowly stitch the forest back together.

The radical aboveground changes wrought by fire are apparent to anyone who witnesses a burnt forest. Less obvious are the subterranean effects on those fungi unable to withstand fire. Fire affects fungi in complex ways—either directly by killing them; or indirectly by altering or destroying their habitats, food sources, and relationships. Fungi, like all organisms, have varying tolerances to temperature. The heating of soil caused by fire manifests as a gradient of death. Most mycorrhizal fungi live in the top 4 inches of soil. Scorching temperatures of high-intensity fires kill fungi near the surface, but those living further down have a greater chance of survival.

In the deeper horizons of the soil, this "backup" of fungi, along with some invertebrates, may shelter from fire. Afterwards, they are thought to move up through the soil profile and recolonize the deserted upper horizon. But they can't do it alone. When trees and other vegetation are lost to fire, habitats are radically transformed, and mycorrhizal fungi will perish without their plant partners. If the plant dies, it's a death warrant for the fungi. Some mycorrhizal fungi might cling to life for a while by tapping into the stored sugars in the stumps of burnt trees, but unless they can hook into another mycorrhizal network, they too will perish.

More Tricks and Tactics for Surviving Fire

Other fungi have surprising fire-survival tricks. Some truffles in the genus *Mesophelia* produce new post-fire odors. Usually smelling more subtly nut-like, once scorched they stink of rotting onion. This makes them more detectable or desirable to mammals that excavate and eat them, then expel the fungus spores elsewhere in their feces. Fungi such as native bread consolidate their hyphae into a food supply called a sclerotium that requires fire to rouse it from dormancy. Other pyrophilous fungi don't adopt any of these strategies, yet somehow manage to survive. The big mystery is where and how they endure the extended periods, sometimes decades, between fires.

Combining talents by forming relationships is another good strategy for coping with extremes. Mycologist Patrick Matheny and his colleagues suggest that some pyrophilous fungi hide out in the tissues of lichens, club mosses, and bryophytes. Their theory is known as the "body snatchers hypothesis." These fungi don't just inhabit the surfaces of their hosts but are harbored within their cells. Plants can't flee from fire, and many are too large and exposed to avoid being burnt. But lichens, club mosses, and bryophytes are small, and can colonize tiny crevices and cracks in rocks. These small-scale refugia—havens where organisms wait out unfavorable conditions—allow them and the fungi they harbor to withstand fire. When the host is exposed to fire, the fungus becomes reproductively active.

The theory sounds plausible for less severe fires, but how these organisms withstand more severe fires is not known. One idea is that new lichen fragments may simply blow in from unburnt areas after fires. If a lichen fragment lands in a suitable spot, it can potentially regenerate vegetatively. Its pyrophilous fungal cargo can then colonize the new lichens or inoculate soils. When the fungus reproduces, its airborne spores blow onto other regenerating lichens and plants, germinate and take up residence in these new hosts.

Most of this takes place at a scale too small to see, and the relationships between fungi and their hosts are only just becoming understood. While hiding between fires, pyrophilous fungi might also tap their hosts for nutrients. The fungus benefits from food and lodgings, but what's in it for the host? The give and take of relationships is not always straightforward, but some researchers suggest the fungus might return the favor by supplying mycotoxins such as alkaloids. These chemicals increase the host's resistance to stress and make them less palatable to hungry herbivores.

The Old-Forest Advantage

Forests are at their most vulnerable following fire because soils and any remaining life they contain are exposed. With the loss of forest canopy and the protective layer of leaf litter, the forest floor has little buffer from sun and wind, rain, frost, and snow. Increased penetration of sunlight raises the soil temperature and reduces moisture. Soils lose structure, becoming unstable and prone to being blown or washed away. When organic matter is lost, so too are the habitats and food supply of saprotrophic fungi that in turn supply nutrients to plants.

Loss of organic matter in soils is directly related to fire intensity. If the fire is relatively hot, the organic matter combusts, usually leaving only highly alkaline ash in which few organisms can survive. Less intense fire causes only the partial combustion of organic matter but changes soil chemistry, creating water-repellent soil layers and preventing water from infiltrating.

The complex combination of direct and indirect effects of fire on fungi play out in different ways. Every fire is different. It depends on when a fire occurs, its intensity and patchiness, the passage of time since the last fire, and the age and condition of the habitat. Severe fire usually radically reduces the abundance and diversity of fungi. The more extreme the changes to their habitats, the longer it takes them to recover. Older forests have more complexity and diversity, and hence more resilience to fire and other disturbances. Fungi in older forests recover faster than those in younger ones because of their biological legacies—the logs and stags, organisms and propagules, and other living and dead components that remain after a fire. Large aggregates of soil and dense mycelial mats also sometimes prevail. Logs and remnant organic matter provide stores of spores and seeds, and other reserves to kickstart recovery. They serve as the biological lifebuoys of the forest: retaining moisture, limiting soil erosion and the loss of carbon and nutrients, and speeding up colonization and recovery. However, as the world's forests become younger with the felling of old growth, forests and their fungi face greater threats from fire.

Pyrophilous fungi may be adapted to fire, but other fungi depend on the absence of fire for their survival. While it's not entirely known whether these fungi are intolerant to fire per se, they prefer the conditions in forests that have not burnt. They typically inhabit well-decayed logs in cool, moist forests that are usually old and undisturbed. The closed canopies; and old logs, deep litter, and abundant soil nutrients of these forests create conditions favorable to many fungi. Thick and insulative moss layers accumulate over decades, keeping logs moist and cool, even during drought. One of Australia's rarest fungi, teatree fingers, has only been found in long-unburnt habitats. Rare fungi are usually so because their hosts, partners, or habitats are rare.

As the world heats up and dries out, fewer forests are untouched by fire. Fire is not only becoming more widespread, but it is penetrating ecosystems such as rainforests that are not adapted to cope with fire. This has mycologists worried, as fire is spreading faster than the ability of these ecosystems to adapt, throwing the long-term survival of their fungi into question.

The Lonely *Entoloma*

Pam Catcheside is a mycologist who has been tracking how fungi respond to fire. In early fall 2010, Pam and a group of naturalists were surveying fungi in the Ravine des Casoars on Karta Pintingga (Kangaroo Island), off the South Australian coast. They were keen to see what fungi might be found following a fire three years earlier. One of the surveyors, Helen Vonow, came across a collection of tiny mushrooms on a piece of fallen bark from a sugar gum tree. It first appeared to be a kidney-shaped species known as the variable oysterling. On closer inspection, Pam hesitated and took out her loupe. Her keen eyes detected subtle variations in the pink coloration of its lamellae, hinting at another identity. "I examined it under the microscope that evening and its angular—rather than globe-like—spores loomed into focus," she explained. The signature spores confirmed her hunch. They had discovered a new and rare species. Now known as *Entoloma ravinense*, or the hidden pinkgill, it has been classified as endangered under the criteria of the International Union for Conservation of Nature (IUCN).

Entoloma is a large genus of diverse mushrooms that are found all over the world. They're commonly known as pinkgills in reference to the color of their lamellae. Many are rather ordinary or drab, while some, such as the blue pinkgill, are spectacularly beautiful. Some form relationships with trees while others are recyclers. Some are edible, others poisonous. Several are rare and appear on the IUCN Red List of Threatened Species. Many are notoriously difficult to identify, unless you're Pam.

The hidden pinkgill grows in the narrow geographic range of the western end of Karta Pintingga and only on sugar gum. This tree is endemic to the state of South Australia, growing only in a handful of places, but has naturalized in other countries. Stands of sugar gum are also widely planted in Australia for firewood. However, life is complex, and specific conditions and influences collectively determine the niche of a species. For the hidden pinkgill, it seems that not all bark is equal. It appears to favor the underside of bark, and only that which has been on the ground for a while and is nicely rotted. It also seems partial to large sheets of bark. Sugar gums shed their bark after fire and hence fire is likely to play a role in providing habitat for the fungus.

Karta Pintingga harbors a significant area of natural remnant vegetation, including many endemic plants. More than a third of the island is designated as national park or protected wilderness area, including the Ravine des Casoars. Pam has recorded an astonishing 750 fungus species among the island's habitats, and she is convinced many more remain to be discovered. In the summer of 2019–20, fire hit again. The high-intensity blaze ripped across the island and ravaged much of its native vegetation. Most of the western end of the island, including 96 percent of Ravine des Casoars, succumbed to the inferno, which took a fortnight to extinguish.

Local ecologies, wildlife, and human residents all suffered a dreadful battering. But what of the fungi?

A few months after the fires, Pam and the group returned to search for the hidden pinkgill but with no success. So much remains unknown about how it survives with what appears to be very particular requirements. How hot is too hot for its mycelium and spores? Was the fire too extensive for the limited population to survive? Could the bark of other tree species be suitable, or will it only colonize sugar gum? After disturbance, a fungus needs a while to settle into its new home and go about producing mushrooms. "It seems to take about three years for this species to fruit, so there's hope it might still appear," Pam told me. We talked about what these rare species tell us about the broader ecology of the forest and what their loss could mean. Different fungi perform different processes, and the greater the diversity of species, the better a forest functions and the more able it is to withstand stress. We wondered how many other fungi occupy particular niches and play specific roles in supporting the forest's ecology.

Pam's profound sadness following the fires was palpable. It was not just for the losses experienced by her friends and other islanders, but also the radical destruction of habitat. She was heartened to see the vegetation recovering, but it will take longer to know how the hidden pinkgill fared. "Poor little thing," she said. "It's terribly rare and only found in two locations, and I'm very concerned about its fate." Pam is ever pragmatic. She understands the vital role of fire in Australian ecologies. Although fire is destructive, perhaps the hidden pinkgill could rely on it. At this point, though, the fate of the newly described species remains a mystery. Its short stint in the limelight prompts the question: how many other species might disappear before they are named or even noticed?

Fungi in the Pyrocene

Climate change is radically altering the nature of wildfire in many parts of the world. Fires are becoming more frequent and intense, and less predictable. The "fire season" has grown longer, and fires have morphed from megafires to gigafires and beyond, threatening a recalibration of scale. Our current age is popularly described as the Anthropocene, the human epoch. The term was coined by meteorologist Paul Crutzen and biologist Eugene Stoermer in 2000, and reflects the human-induced changes to the earth and atmosphere. Fire historian Stephen Pyne took the paradigm further, coining the term "Pyrocene" in 2015 to describe our current "Fire Age." As Pyne puts it, "Its core premise is that we made an alliance with fire that gave us small guts and big heads, and then took us to the top of the food chain, and now threatens to unhinge the planet." Highly destructive fires will become more common in places like Australia, the western United States, parts of the Mediterranean, and central Europe, as climate change causes wilder weather, soils and forests to dry out, and droughts to lengthen. As Pyne reminds us, "Even Greenland is burning."

The concept of fire as a management tool is highly divisive. Some scientists, conservationists, and others are at odds with land managers about whether fighting fire with fire makes sense. Those interested in diverse and resilient forests don't always agree with burning regimes that prioritize, for example, an economic resource such as large trees destined for felling. In this scenario, frequent low-intensity fires are used to remove what is termed "fuel" or "hazardous fuel"—combustible organic matter—with the aim of protecting larger trees. The term "fuel" is loaded, as it carries the connotation of being the cause or part of the cause of the problem. "Fuel" is a contributor, not a cause, of fire. There is no single cause, but a web of interacting relationships, processes, and feedbacks, all catalyzed by a changing climate. Seen through another lens, "fuel" is essential food and habitat to fungi that retains moisture, stabilizes soil, and supports the growth and resilience of the same trees destined for felling.

Fire studies suggest that although fire initially reduces the richness of fungus species in the years immediately following, many soil fungi are resilient and can recover from fire given enough time. However, the required long timeframes usually don't align with those of forest management. Some fungi need decades to return, and this relies on their habitats not being subsequently disturbed. Given the increasing use of fire and the shorter intervals between these burns, the chance of fungi not being "subsequently disturbed" is unlikely. Some scientists question whether current burn regimes allow sufficient time for ecosystems to rejuvenate. Ecologist Elle Bowd and her colleagues describe the shock finding of their research on the effects of disturbance caused by fire on the soils of Australian mountain ash forests. They found decreases in key soil nutrients an astonishing 80 years after fire. These deficiencies could persist longer, especially in forests with histories of burning or logging. The researchers' work reveals that land managers could be grossly underestimating the timeframes needed for forest recovery following burns. It's also a reminder of how much more needs to be known about fungi in order to connect the general patterns that Elle and her colleagues found to the responses of individual fungus species.

The shortening of intervals between prescribed burns also changes the nature of ecosystems, altering forest structure and increasing their vulnerability to further fire. One size does not fit all when making assumptions about fire, especially when it comes to the effects on less understood organisms like fungi. A burning regime that might be effective in one ecosystem type, with its localized climate, fire history, and ecology, could devastate another. In southern Australia, repeated prescribed burns have converted rich and resilient wet forest ecosystems into flammable scrublands. The complexity of fire and ecosystems warrants a more considered approach than simply rolling out frequent prescribed burns, which some scientists consider to be at least as ecologically destructive as larger fires.

Prescribed burning could perpetuate the problems it aims to mitigate—leading to a loss of species, ecosystem complexity, and resilience. It also escalates other threats. Burnt landscapes make easy hunting grounds for feral predators that prey on surviving mycophagous (fungus-eating) mammals. Without their mammalian vectors, truffle fungi cannot disperse. Feral herbivores, such as deer that nip off new shoots of plants, can also hinder recovery of ecosystems after fire.

Protocols and policies for fire seldom explicitly include fungi. Most often they are premised on vegetation communities, "fuel loads," and endangered animal or plants. Many fungi are likely to be swept along by more general efforts to protect habitats. However, some fungi have specific needs and risk being overlooked. Given that the great range of organic matter that fungi inhabit is only seen as "fuel," they are likely to be left out. The popular catchcry that fire is a "natural part of the landscape" may well be true, but how much burning is too much? Is enough known about varied and complex ecologies to use fire productively and not destructively?

Indigenous Australians have used fire for thousands of years. They walk slowly over Country, observing and lighting low-intensity fires. It's an entirely different approach from that used in parts of northern and western Australia and elsewhere in the world, which involves throwing incendiaries from helicopters. While the mechanization of fire management claims to manage fires more "efficiently," it operates at a different scale. It's hard to notice from a helicopter the response of a skink or a fungus to a fire. The nuance of many small losses may go unnoticed.

In the spring of 2006, following the Warrinaburb fire, I returned to Gariwerd. With a 360-degree view from the summit of Duwul (Mount William), the extent of the fire was apparent, as was the extent of recovery. The fire had ripped through the tree crowns, killing many, but some survived. Along the Wannon River, blackened trunks sported luminous new green leaves, while others remained lifeless. On the ground, trails of ants ferried cargo among orchids and the "open mouths" of banksia cones that had surrendered their fire-triggered seeds. The majestic flower spikes of grass trees were alive with the buzz of insects. As I climbed higher, I spotted the convoluted caps of morels, which seemed to mimic the weathered rock formations above. Each life form offered hope among the devastation, but I wondered whether the diversity of species would return before the next fire struck, or ever.

Climate change plays out in many ways. As parts of the world burn, others melt. For COP26, the 2021 United Nations Climate Change Conference, the foreign affairs minister of the South Pacific island nation of Tuvalu, Simon Kofe, recorded a speech while standing in the ocean. Dressed in suit and tie and with the sea water lapping around his knees, he reminded the world that climate change is not something of the future. It is clearly happening now. Climate change effects do not manifest evenly across the world. They are felt differently and in varying degrees depending on latitude, altitude, and a suite of other influences. Low-lying nations are especially vulnerable to rising sea levels.

The changing climate is also felt more acutely at high elevations. Alpine species react to heating by clambering ever higher in search of their ideal living conditions. The question remains as to whether they can keep pace with climate change, or if some species will get left behind. Switzerland is one of the most mountainous countries in the world, with almost two-thirds of its land area made up of mountains. I headed to Europe's longest glacier, the Aletsch Glacier in Switzerland's Bernese Alps, to find out how fungi respond to the thawing ice.

Melting Biodiversity Hotspots

I feel a sharp tug on my climbing harness, hear a cry for help, and turn around to see a head in the snow. One of my companions has slipped into a crevasse. The nine yards of rope between us keeps him from plummeting deeper into its icy depths. It takes our guide, mountaineer Hansueli Klossner, only minutes to hoick him out. There's a collective shudder of relief. Navigating glaciers has become trickier in a heating climate. They change more quickly, more radically, and in unexpected ways. I peer into the deep blue abyss of the crevasse and hear the eerie creak and groan of the glacier on its relentless downhill march.

We pass the tributary glacier that sits below the Fiescherhorn summit and clamber along a moraine. Rocky boulders sit suspended on plinths of ice. Weather is a whimsical sculptor. Traversing the glacier is an intense experience. It's not only the sense of one's own insignificance it commands, but the disquieting realization of its transience. My hiking companions and I marvel at the elemental landscape surrounding us, but I sense that they too grapple with a bittersweet snarl of awe and angst.

The high-pitched whistles of alpine choughs sound overhead. They're an inauspicious reminder that high-altitude species are more prone to environmental changes because they live closer to their physiological limits than those at lower elevations. Since 1970, the climate in the Swiss Alps has warmed by about 2.7 degrees Fahrenheit. With the climate changing, species that live on the summits of mountains progressively lose habitat. As they migrate upwards, there is less available territory and more competition to inhabit it.

The weather starts to close in, and the crunch of crampons on ice picks up speed. We finally reach the vertical rock face that rises over the Konkordiaplatz where three glaciers merge. Above the rock face is our destination for the night, the Konkordia Hut. Built in 1877, it sits at an altitude of 9,350 feet above sea level. Unless you can hitch a ride in a helicopter, the only way to access it is across the ice and up the rock face. Back in the nineteenth century, the hut was 160 feet above the glacier and hikers scrambled up the rocky terrain to reach it. Over time, the glacier has melted, receding not just in length but also in depth. As the glacier's sinking accelerated, ropes were strung, then wooden ladders were erected to help hikers scale the growing rock face to the hut. These rickety structures were updated in 1970 when steel staircases were bolted to the rock. I unclip my crampons, step from ice to steel, and begin the 467-step ascent.

Today the Konkordia Hut sits 650 feet above the glacier. Each year the staircase is extended by a few steps as the glacier subsides. With mountains becoming less stable and rockfalls more frequent, climbing the steps becomes more precarious. The hut may eventually have to be abandoned or relocated. The conspicuous zigzag of steel steps is a stark visual reminder of the changes wrought by global heating. Less visible are the changes playing out among mycelia beneath the sliver of alpine soil.

The Alps are recognized as a biodiversity hotspot, with distinct niches for specialist species including fungi. But hotspots are literally getting hotter, with climate heating in alpine areas happening faster than below. Scientists are documenting these changes to understand whether species can respond and adapt accordingly. Noticing seasonal and cyclical phenomena around us is an old art. It's how we read environments. All animals do this to survive. Scientists refer to this seasonal timing of biological events in an organism's life, such as breeding, as phenology. The availability of resources for a fungus, such as food and water, for example, is tightly linked with the timing of phenological events. Climate change, particularly temperature increase, is expected to profoundly affect the phenology of fungi. I headed further east, to the Swiss town of Davos in canton Graubünden, where a team of scientists is studying the responses of biodiversity to climate change.

Seeking Cool in a Heating World

The alpine town of Davos sits at 5,120 feet above sea level. In the nineteenth century it was a recognized *Luftkurort*, literally, "air curing place," where sufferers of bronchial and other ailments came to recuperate and inhale the clean mountain air. It has long been a ski resort but today is better known for hosting the annual meeting of the World Economic Forum. Many of its thousands of attendees—political leaders, investors, businesspeople, and economists—converge on and perhaps inhale the air of the "Golden Egg," the luxury egg-shaped AlpenGold Hotel with its shining shell of golden steel. While talk of transitioning to a greener and more equal global economy is on the agenda, their hundreds of private jets jostle for space on the tarmac.

Should there be a landslide, and the Egg be dislodged and roll down the hill, it could well slam into the Institute for Snow and Avalanche Research (SLF), where researchers are tracking climate change. The institute's scientists research and monitor long-term changes in forests and landscapes, natural hazards, snow, ice, and climate. As individual species have differing traits, ecologies, and distributions, they respond in different ways. The Alps are ideal for studying the responses of species to climate change because of the region's long history of research, with the first experimental research plots set up in 1888. The institute is a branch of the tiny nation's Federal institute for Forest, Snow and Landscape Research (WSL), which employs 600 research professionals.

Yann Vitasse is a WSL forest ecologist who, together with his team, examined the data of over 2,000 species of alpine animals, plants, and fungi to assess how and if they are adapting to a changing climate. The researchers found that different groups of organisms are moving upslope to cooler latitudes and altitudes, but at different rates. Woody plants, for example, migrate upwards at a faster rate than aquatic insects, birds, ferns, or wood-decay fungi. Some species could be favored by climate change while others may disappear. Species that can disperse themselves more easily and those with shorter life cycles may be more responsive to change. While changes to land use explain some shifts, the consistent upward movement across almost all the species studied is likely to reflect the increasing temperature and receding snow cover resulting from climate change. Most worrying was the researchers' conclusion that the upward migration of fungi, plants, and most animals is too slow to keep pace with temperature shifts.

As the climate changes, so do the seasons. With snow melting earlier, spring gets a head start. Many species not only migrate upslope but respond by bringing forward the timing of their spring activities such as reproduction. Other species, however, don't respond, or respond less, to the changed seasons. These differences in the rates of response mean that species may not be able to coordinate their activities with each other. For example, predators rely on prey species emerging at certain times. Plants flower to coincide with pollinators. Fungi also respond to the phenology of both animals and plants, especially those that are partners or vectors. Because the lives of fungi, plants, and animals are intimately intertwined, a change in the phenology of one can bring about a cascade of consequences for others.

Almost all research on the effects of climate change on alpine species has focused on animals and plants. However, a study by US ecologist Jeff Diez and colleagues recorded the altitudinal changes in fungi in the Alps. By looking at historical records they found that more fungus species are producing mushrooms at higher elevations in response to climate change.

In general, the "mushroom season" in many parts of Europe is starting earlier each year and extending longer, which might sound like a boon for mushroom foragers. However, hidden by the generalizations is a more nuanced interplay of local climates and conditions, and the specialized requirements of different fungi. For example, in recent decades, the witches' cauldron—a fungus that typically grows in moist, nutrient-rich habitats adjacent to waterways—has increased in abundance in some European countries but become locally extinct in others. Researchers in Finland attribute its increased abundance in their country to milder winters, wetter springs, and changes in seasonal flooding patterns—trends all associated with climate change. Climate change–induced drying in other countries could account for its disappearance there. These observed shifts could reflect changes in fungal processes below ground, potentially affecting food webs, nutrient cycling, and the timing of nutrient availability in ecosystems.

Another group of SLF researchers, led by soil ecologist Frank Hagedorn, experimentally tested the response of fungi to climate change by artificially heating soil. They found that increased soil temperature changed the composition of the fungus community, with different species responding in different ways. The milking bonnet and the rufous milkcap, for example, both increased in abundance. The splendid woodwax almost completely disappeared. Other fungi showed no response to the temperature change. Conducting scientific trials in the field is difficult, as it can be tricky to isolate variables and decipher the effects of climate change against natural background changes. Altering one variable also causes changes in others, so it is hard to know to which variable a species is responding. Heating the soil in Hagedorn's study, for example, increased microbial activity. Increased microbial activity enhanced the decomposition of humus and doubled the amount of nitrogen it released. As the rufous milkcap is a nitrogen-favoring species, its increased abundance is thought to be in response to the increased availability of nitrogen, rather than to increased temperature itself. The study highlights the complexities of systems and interactions, and the difficulties in interpreting changes.

As fungi and other organisms march upwards, the alpine terrains become more hostile and elemental, and soils recede. Fungi, plants, and other organisms all need soil. Lichens and other fungi help create soils but it doesn't happen overnight. It typically takes centuries or millennia, so soil formation lags a long way behind the speed of climate change. The lack of soils or nutrient-rich soils in alpine areas will favor species that are able to manage with little soil. Those that cannot will be left behind. As a result, the composition of species in alpine communities is likely to become less diverse and ecosystems less resilient.

Mountainous countries become less stable as the climate changes. Avalanches and landslides are increasing, and Switzerland is urgently trying to keep its snow from sliding down. From the shore of Lake Davos, I look toward the ridge of the Weissfluhjoch. It's striped with avalanche-protection barriers. More than 300 miles of these barriers exist across the Swiss Alps. There are also 186 automatic measuring stations that record weather conditions around the clock. They transmit hourly data to scientists who produce twice-daily bulletins about avalanche risk and have done so since 1945.

I wander from the lake to the institute and am greeted warmly by alpine scientist Frank Graf. Frank was hosting me to give a talk on Australian fungi, but I was keen to hear about his research investigating bioengineering for erosion control and slope stabilization in the Alps. In a research project conducted on erosion-prone slopes in nearby St Antönien, Frank and his team showed how vast hyphal networks of mycorrhizal fungi help mitigate the effects of climate change by stabilizing landscapes. Fungi provide mechanical stability to soils, assembling soil particles into aggregates, then bonding them together with sticky cementing substances called polysaccharides. "One of the main requirements of eco-engineering and, actually, any soil restoration, should be to establish conditions favorable to the formation of stable soil aggregates," Frank explains. "These aggregates help soils retain water and nutrients that are needed for plant growth, and protect them from leaching." Fungi also contribute to stability by fostering the development of a cover of protective vegetation, and accelerating and increasing the growth of plant roots. Without an intact mycorrhizal foundation, slopes that are prone to erosion are more likely to collapse. However, when soil is highly degraded due to practices such as agriculture, it is harder for mycorrhizal fungi and their plant partners to contribute to soil stability. Says Frank, "When soils are badly affected by erosion, even pioneer plants reach their limits." Forests, fungi, and intact soils all go hand in hand.

Fungi are versatile. Many respond and adapt to changes in their environments, but there's always a limit. Across the world, scientists race to come up with "solutions" to the effects of global heating. In Australia, small fires were traditionally used to prevent infernos. In Switzerland, explosives have long been used to trigger small avalanches to prevent catastrophic ones. Both practices are likely as old or older than anthropogenic global heating. However, their scaling up in reaction to the effects of global heating is becoming less effective in reducing disastrous fires or avalanches, instead causing cascades of new problems. As the planet heats, scientists seek new innovations that go beyond upscaling to focus on causes. Although research largely investigates aboveground ecologies, their survival hinges on a functioning subterrain. Networks of mycorrhizal fungi are a major carbon sink in terrestrial ecosystems. Reducing the vast and seemingly endless disruption of soils to maintain a healthy carbon sink could be a vital starting point to reducing global heating.

In 2021, evolutionary biologist Toby Kiers launched an organization called the Society for the Protection of Underground Networks (SPUN). Toby tells me it is as an "underground climate movement" that aims to put fungi on the conservation agenda by protecting fungal networks. SPUN works in collaboration with scientists, advocates, and local communities across the globe to map fungal networks and quantify biodiversity hotspots. Toby emphasizes the need for a shift in conservation thinking that moves beyond focusing on high-profile species—usually animals and plants—to protect whole systems. She refers to mycorrhizal fungal networks as an "ancient underground life-support system." Protecting them begins with mapping and determining their location, especially highly diverse biodiverse networks and those that sequester the most carbon. The team aims to understand how carbon flows through soils, and what stimulates fungi to draw carbon down deeper into their networks. This will allow SPUN to monitor shifts in response to climate change and determine which fungi can withstand increased soil temperatures and drought.

Ideally, policy makers could then prioritize areas based on networks that sequester the most carbon and have the greatest diversity. To tackle global heating, says Toby, we must address the "global blindspot of vast underground fungal networks that sequester carbon and sustain much of life on earth." It's a bold step toward the many ways of thinking and innovative approaches that are needed if we are to cool our heating planet.

6

—

FUNGAL
RENEGADES

In a lifetime of adventuring and filmmaking, eminent biologist and broadcaster Sir David Attenborough has revealed some of the world's most extraordinary life forms. From the flamboyant plumage and outlandish behaviors of birds to the mimicry and chicanery of plants and insects, he has captivated audiences with the bizarre forms and strange behaviors of the natural world. Among them are some of the most beguiling fungi of all—those that glow in the dark, and those that invade and commandeer the bodies of other organisms.

Given most fungi are microscopic and many others are small, brown, and nondescript, it's hardly surprising that fungi with peculiar forms, conspicuous colors, or curious habits capture attention. The great array of different fungal forms are thought to have evolved as strategies for ensuring and maximizing spore dispersal. The umbrella mushroom form protects spores from rain and hoists spore-bearing surfaces above the ground to benefit from tiny currents of spore-dispersing wind. For puffballs, it's a numbers game, with some larger ones estimated to contain trillions of spores. However, why a fungus has striking colors or surprising odors, or produces chemicals hallucinogenic to humans, is still being figured out.

Fungi produce an astonishing range of chemicals. Fungal toxins occur in species across the kingdom and are thought to have independently evolved several times throughout evolutionary history. Although it's common to hear that a mushroom "looks poisonous," the notion of warning coloration—or aposematic coloration—is probably a misconception, at least when it comes to fungi and humans. Some lethal fungi have drab and understated colors, while some of those that are red or yellow—colors often perceived to be warning colors among animals and plants—are harmless to humans. Although we don't yet fully understand the complex cocktail of chemicals produced by fungi, noticing their colors, odors, and peculiar habits not only prompts important questions for science but delights the senses.

Glowing Ghosts

One moonless night in early fall, I wander along the beach at Rossiter Bay in Australia's south-western corner. As the world becomes ever more flooded with artificial light, even here in the vast state of Western Australia, I relish the nocturnal respite of darkness. Sounds and scents intensify as the night gradually grows darker and my vision dwindles. A mob of kangaroos thumps through the sand, followed by the agitated squeaking of a pied oystercatcher. I tune in to the gurgle of water draining from between the rocks and the clacking of scurrying crabs. Mingling scents of wet sand, decaying seaweed, and damp coastal vegetation waft into my nostrils. Gradually the endless expanse of night sky fills with stars.

As hunger kicks in, I head back to camp through the coastal scrub. Feeling for the sandy trail with my bare feet, I shuffle along, wending my way between faintly silhouetted outcrops of granite and gneiss. Insect wings flicker against my face as I brush against vegetation. I apologize to an invisible bird that startles with a squawk and flaps away. Tripping on an exposed root, I land face first in the sand. It is here, from my unexpected vantage point, that I spy a faint glow through the undergrowth. I squint and try to focus through the blackness. Are my eyes playing tricks on me? Or could it be dappled moonlight? I glance up. There's no moon. Scrambling to my feet, I push through the tea-tree toward the glow and find a perfect cluster of overlapping ghost fungi. Had I a torch or had there been moonlight, and had I not tripped, I might have passed them by.

The glow of the ghost fungus is at first subtle to detect, and your eyes need to fully adjust to the darkness. Edging closer, I count eleven mushrooms clustered together at the base of a tree trunk. There are probably more hidden beneath. One is the size of a dinner plate. Others in various stages of growth have inverted into funnels. I sit and marvel at them as their luminescence appears to intensify to a faintly and eerily green glow, making the dark coastal scrub a hauntingly beautiful place. Despite having witnessed this enigmatic fungus dozens of times, the otherworldly thrill I experience has never waned.

The ghost fungus is conspicuous for its bioluminescence (*bios* meaning "living" and *lumen* meaning "light")—a biological process whereby light is emitted by living organisms. Although we can't detect its bioluminescence in daylight, the ghost fungus glows continuously. But even without its magical glow, it is obvious because of its baseball-glove size and the dramatic clusters it forms. It is thus a long-known species and was one of the first agarics (fungi with lamellae) in Australia to be scientifically named. In 1841, it had caught the eye of Scottish naturalist James Drummond, who collected a specimen growing on banksia wood further north along the Swan River.

Endemic to Australia, the ghost fungus grows in a range of habitats, from eucalypt forests to pine plantations, alpine foothills, and coastal scrub. Although common, it appears in many guises, and exhibits an astonishing array of colors. Young ones can be matte bluish-black. Others are creamy white and streaked with orange or golden tones. Sometimes they're reddish-brown. Occasionally they're purply mauve or even greenish. Most often they are white, variously splotched with other colors, particularly toward the center of the cap. As they age, the funnel-shaped mushrooms buckle and contort, scalloping along their edges like the hem of a dancer's flowing dress. This fungus is a superb example of how the form and features of fungi can vary depending on their developmental stage and on the environmental conditions to which they are exposed. Its kaleidoscope of colors and mixed identity kept early mycologists on their taxonomic toes. Being something of a chameleon, the ghost fungus has been renamed at least thirteen times since 1844, when British naturalist Miles Berkeley scientifically described Drummond's specimen. Since 1994, the name *Omphalotus nidiformis* seems to have stuck.

Bioluminescent organisms have long been noticed, and have inspired—and, at times, terrified—observers. Historical observations of bioluminescence appear in ancient Chinese poetry, and fireflies and glow worms feature in age-old Buddhist texts. The scientific study of bioluminescence dates to the first written records of the ancient Greeks, who noted marine phosphorescence in 500 BCE. Aristotle (384–322 BCE) described marine bioluminescence as "exhalations of fire from the sea." He was among the first to note terrestrial bioluminescence in the phenomenon of "glowing wood" or "shining wood." The glow is emitted by luminescent mycelia within decomposing wood, although this association wasn't established until the nineteenth century. Glowing wood has attracted other names, including fairy sparks, fairy fire, cold fire, foxfire, and torch wood. English philosopher and scientist Francis Bacon (1561–1626) performed some of the first experiments to determine the nature of shining wood. His observations that wood must be moist and that "putrefaction spreadeth" provided a vital clue for future researchers.

It was deep in the coalmines of Bochum in the Ruhr Valley of Germany that the fungal origins of shining wood became apparent. The director of the Board of Mines noticed that the luminescence in supportive mine beams occurred in streaks. The streaks were in fact luminous "fungal threads," and their luminosity was bright enough for miners to work without lamps. The significance of these findings, published by the German botanist Theodor Nees von Esenbeck in a scientific paper in 1823, was that luminescence was caused by a living organism, not a chemical process. As with many novel ideas in science, it was met with skepticism. It was only accepted in the 1850s, after Austrian chemist Johann Florian Heller examined the shining wood under the microscope and confirmed that light was being emitted by "fungal threads." He called the fungus *Rhizomorpha noctiluca*.

However, in the thousands of years before these revelatory discoveries, bioluminescent fungi likely appeared in the stories of Indigenous Australians and other indigenous peoples. Early colonists in Australia recorded Aboriginal reactions to what is thought to have been the ghost fungus. Aboriginal peoples respond to fungi in different ways, with some using them to their benefit and others fearing them. Drummond recorded the fearful reactions of several Western Australian Aboriginal people, who referred to it as *chinga*, meaning "spirit." Drovers in New South Wales in the 1930s noted the unease of Aboriginal people when camped near luminous fungi. Luminous fungi were observed and inspired similarly mixed responses elsewhere in the world. Some Micronesian peoples destroy luminous fungi, believing them to be an evil omen, while others have used them in body decoration, especially to intimidate enemies. The Dutch botanist Georg Rumphius (1627–1702) recorded Indonesians of the island of Ambon carrying luminous fungi as lanterns at night. In Australia I've asked about ghost fungi when on Country with First Nations peoples, including Wiradjuri, Yorta Yorta, Dja Dja Wurrung, and Wadawurrung, but the people I have spoken with were unsure if these fungi held significance for their communities.

Bioluminescence in bacteria and animals has been well studied, and today is a focus of genetic engineering. Less is known about the nature of bioluminescence in fungi, although recent research is shedding light on their secrets. Two things are usually involved in the bioluminescence process. The ghost fungus contains a light-emitting substance called luciferin (although *lucifer* sounds devilish, it was more likely named for the Latin, which means "light bringing"). In the presence of oxygen, luciferin is oxidized by an enzyme called luciferase. Luciferin and luciferase are not single compounds but refer to the substances and enzymes that vary between bioluminescent organisms. As a result of this chemical reaction, energy is released as a pale greenish light. It was French physiologist Raphaël Dubois who made these discoveries in 1885 using bioluminescent bivalve mollusks and click beetles. Researchers since Dubois have studied the bioluminescence of numerous other organisms, mostly marine species. In the second half of the twentieth century, eight different classes of luciferins were discovered, along with the mechanisms of their light-emitting reactions.

However, none of these are used by fungi. Recently, a team of Russian and Japanese researchers revealed that the luciferin found in fungi is chemically unrelated to those of animals and microbes, representing a different mechanism for emitting light. They found that bioluminescence in many fungi is based on an antioxidant or "luciferin precursor" called hispidin. It is not hispidin alone that enables the fungus to bioluminesce, but its ability to produce two enzymes that are necessary for bioluminescence. Their research has answered one set of questions while opening up a suite of others.

Bioluminescence is a complex phenomenon. It has evolved many times in evolutionary history for different reasons. Scientists have unraveled many of its mysteries in the past three millennia. If you are a deep-sea organism, finding a mate or even trying to communicate presents challenges in the pitch-black watery depths. Likewise, finding a feed can be equally onerous. Viperfish and anglerfish, among others, have evolved dangling, glowing appendages to lure prey within gulping distance. Others use light to ward off predators. Although bioluminescence is well documented among animals, why fungi bioluminesce largely remains a mystery. It might seem that something as dramatic as bioluminescing would have an equally entrancing purpose. While some bioluminescent fungi are known to attract spore-dispersing vectors such as insects, researchers who experimentally tested this hypothesis found it was not the case with the ghost fungus. It seems that bioluminescence is unlikely to provide any selective advantage, but could function to remove by-products of other cellular processes. David Attenborough questions whether it is simply just a "beautiful by-product of evolution with no function ... just a biochemical accident." Those who find these scientific explanations unsatisfying might prefer to stick with the theory of my five-year-old friend Ava Lee, who assured me that "Ghost fungi help wombats find their way through the bush at night."

Although the backstory for the glowing ghosts is yet to be revealed, could it be Australia's most charismatic fungus? The South Australian Forestry Corporation (ForestrySA) seems to think so. It has expanded its commercial operations from the production of sawlogs and woodchips to manufacturing children's squeals of delight. Ghost Mushroom Lane, as it has been dubbed, leads to a commercial pine plantation (suitably spookier at night) near Mount Gambier, on South Australia's Limestone Coast. It is perhaps Australia's most spellbinding example of myco-tourism. Troy Horn, conservation and fire manager with ForestrySA, first organized these night tours of the plantation about five years ago, and 70,000 people have since experienced the glowing forests. "I get really excited taking people to a place we call Green Lantern Gully. You can imagine their excitement on a perfect night, stars above, then seeing the glow of ghost mushrooms for the first time," Troy says. While the thrill of a night-time outing and the idea of ghosts might well be more enthralling to young minds than the fact of them being fungi, it is an imaginative way to inspire interest in these underappreciated organisms. Troy and his small team of fungus enthusiasts hope to expand public interest by offering commercial foraging tours for edible mushrooms, as well as fungal ecology forays in the local native bush.

In the mid-nineteenth century, English-born Australian naturalist George Bennett commented on what was probably the ghost fungus: "to a person unacquainted with this phenomenon the pale, livid, and deadly light emanating from it conveys to him an impression of something supernatural, and often causes no little degree of terror in weak minds or in those willing to believe in supernatural agencies." While the light of the ghost fungus is not deadly, eat one and you may well feel like you want to die. The ghost fungus looks superficially similar to edible oyster mushrooms (*Pleurotus*) and was once classified in the same genus. This likeness, perhaps combined with its tantalizing bioluminescence, have tempted some people to put them in their mouths. Usually, they quickly come back out: the ghost fungus possesses a powerful emetic called illudin that induces a rapid and violent evacuation of one's stomach contents. Unfortunately, no glowing *Homo sapiens* have to date been recorded.

Fungal Assassins and Mummified Grubs

Glowing fungi have humans beguiled, but parasites present a greater challenge. The mere mention of the word *parasite* can send some people into paroxysms of itching and scratching, conjuring odious images of ticks or leeches. In general, humans tend to value, or at least tolerate, organisms considered either charismatic or useful. Those we can eat or pat, or those deemed as pollinators or as indicator species (proxies for the state of an ecosystem) are usually held in higher esteem because their significance to humans is more apparent. Unlike the partners in mutually beneficial symbioses, parasites don't subscribe to reciprocity. However, while they do not directly benefit their hosts, parasites benefit others by their actions, and are vital to forest function. Parasites influence competition and predation, regulating biodiversity as well as nutrient cycling and energy flows. Like animals and plants, some fungi have opted for a parasitic lifestyle.

I was in the depths of north-west Tasmania's Tarkine Forest when I spied a stick that looked suspiciously like it might not be a stick. Going by its size, it was probably more twig than stick, but I didn't think it was a twig either. Its slightly pointed tip and slender stature was the first clue to another identity. Running my fingers down its length I felt its fine sandpaper-like texture hinting at something non-twig. Sizing up the rather emaciated-looking thing, I reckoned it might in fact be a special type of fungus. Beneath the ground, I suspected there might be a caterpillar attached, on which the fungus fed. I carefully excavated the soil around the "twig" with my pocketknife. Brushing the soil away with my hand, I dug a little further and, sure enough, the perfect mummified form of a caterpillar magically materialized.

This unassuming twig-like sporing body of the vegetable caterpillar is easily overlooked. Originally named *Cordyceps robertsii*, it was the first vegetable caterpillar fungus scientifically recorded in Australasia—in New Zealand in 1836—and is probably one of the largest in the world. While many fungi parasitize plants, vegetable caterpillars seek out invertebrate hosts. Despite their common name, they don't just grow on caterpillars but on a great variety of arthropods, as well as the occasional truffle. Known as *Ophiocordyceps robertsii* today, this fungus usually targets moth larvae from the family Hepialidae, such as the Victorian swift moth.

During the larval stage of their lives, caterpillars spend time mostly beneath the soil in the silk-lined shafts of their burrows. Under cover of darkness, they emerge to forage among the leaf litter. Although darkness might spare them from daytime predators, life in the litter presents multiple risks to a caterpillar, not least the chance of encountering fungus spores. Most of the millions of spores a caterpillar is likely to meet do not pose any great threat. Those of *O. robertsii* are another story.

How caterpillars become infected with *O. robertsii* spores is a mystery. Caterpillars breathe through tiny holes called spiracles on their abdomens, so they may unintentionally inhale spores, or perhaps they brush up against spores that have already sprouted hyphae. The hyphae may then dissolve the caterpillar's cuticle (outer layer) with enzymes, allowing the fungus to penetrate its innards. It seems more likely that the caterpillar consumes the spores along with the organic matter on which it feeds but, whichever way the fungus finds its way into the caterpillar, it's a coup de grâce for the luckless creature. Once the fungus is inside, the caterpillar's interior provides the ideal habitat for the mycelium to proliferate. The fungus swiftly colonizes and liquefies the caterpillar's delicate innards via powerful enzymes that pervade the creature's entire body cavity, effectively consuming the caterpillar from the inside out. In the process, the fungus kills the caterpillar and transforms it into something resembling a fungal mummy (known as a sclerotium). Once satiated, the fungus sends its reproductive structure out through the head of the caterpillar and above the soil surface. It releases its spores out of tiny receptacles called perithecia (minute pores in the perithecia, called ostioles, give it the sandpaper-like texture I felt with my hand), and they are dispersed by wind and passing animals that happen to brush against the fungus. And so the cycle begins again.

I turned the segmented, mummified remains of the parasitized caterpillar over in my hand. I hadn't witnessed the process of its demise in the darkness of the subterrain, just the leftovers of a fungus's meal. Mulling over the caterpillar and the unassuming twig-like sporing body, I wondered how this bizarre union might play out at an ecosystem scale.

Vegetable caterpillars are highly specialized fungi. They have evolved a swag of tricks and chemicals that enable them to manipulate the physiology and behavior of their hosts, inspiring scientists and science fiction writers alike. The name "vegetable caterpillar" refers to the mycelium-filled caterpillar at the end of the parasitization process. There is no vegetable involved, but the name probably reflects times when fungi were thought to be plants. Along with other entomopathogenic fungi (those that grow in or on the bodies of insects), they play an important role in regulating populations of insects and other arthropods such as centipedes, spiders, and scorpions. Arthropods, like fungi, are vital to forest function. At times, forest conditions can change in such a way that they favor a particular species or group of arthropods. These changed conditions can result from local disturbances such as fire or forestry, or more global processes such as climate change. Taking advantage of the new favorable conditions, an arthropod species can multiply rapidly.

The increased pressure of a population explosion on forest resources can trigger a slew of effects. These can deplete resources for other forest inhabitants, and dramatically alter forest dynamics. This is where the parasitic nature of vegetable caterpillars can do the forest a favor. Most species of vegetable caterpillar have limited host ranges, meaning they only associate with a small number of species. For example, one might target a particular ant genus; another might be restricted to a particular beetle genus. This specificity means they are likely to play a role in regulating arthropod population dynamics. By preventing any one genus or species of arthropod from gaining the upper hand, they help keep ecosystems stable. Few insect orders have been spared, with vegetable caterpillars capable of infecting the majority.

Several hundred species of vegetable caterpillars have been described worldwide, occupying diverse habitats from rainforests to alpine environments and deserts. In forests they are found in soil, leaf litter, the canopy, and almost everywhere in between. Undoubtedly the most valued by *Homo sapiens* is the Chinese vegetable caterpillar (*Ophiocordyceps sinensis*) that grows in the alpine grasslands of the Tibetan Plateau and the Himalayas. It has been harvested for centuries and is used today in vast quantities for traditional medicine, revered for its multiple apparent health benefits, and its restorative and tonic effects.

The best-studied group of vegetable caterpillars, however, are those that parasitize ants. Much has been written about the so-called "zombie-ant fungi" of the *O. unilateralis* clade, which can modify ant behavior. This species penetrates the ant's cuticle, infiltrating its body, invading and commandeering its muscles. The ant effectively becomes a prisoner in its own body as the fungus swiftly takes the reins and compels it to climb a plant stem. Here the fungus releases chemicals that direct the contraction of the ant's jaw muscles, forcing it to latch on to the underside of a leaf. With its body cavity now flooded with the mycelium of the fungus, the ant dies. From this elevated vantage point, the fungus sends it spore-laden reproductive structure out through the head of the ant, raining spores on its unlucky ant allies below. With their rather ghoulish means of survival, it is little wonder vegetable caterpillars attract the macabre fascination of zombie afficionadi.

Returning to *O. robertsii*, this fungus holds a special significance in New Zealand. Known as *āwheto* to Māori, they were once collected in large quantities. The fungi were hung up to dry and then burnt to produce a charcoal. This was then ground into a powder, mixed with bird fat, and used as a pigment for *tā moko* or facial tattooing. Although the fungus does not look particularly appetizing, Māori sometimes ate it as food, but probably only during times of hardship. It was also used medicinally as an antibiotic and in the treatment of asthma. Lying on the forest floor as I photographed this curious fungus, my focus shifted outwards. I realized I was surrounded by them. It seems I was in the middle of a fungus feeding frenzy. I watched as a moth flew between the legs of my tripod. At least one caterpillar had escaped these fungal assassins.

The Lobster Enigma

Cordyceps attack invertebrates; other parasitic fungi invade plants. Then there are the mycoparasites—fungi that attack or enslave their own. It was in the Cascade Mountains of America's Pacific Northwest that I met with one of the more astonishing mycoparasites.

I knew biologist Sallie Tucker Jones was a patient person

when I found her in the darkened arrivals hall of the Portland International Airport in Oregon. She'd been waiting for me for two and a half hours, during which time, unbeknown to Sallie, I was being interrogated by customs officers in a windowless back room. My apparent crime was my intention to partake in a mycological convention. Three officers tapped furiously on computers while grilling me about "mike-ology," who these "mike-oh" people were, and how many "mike-oh" things were hidden in my bag. It would have made for a hilarious comedy skit were I not the central protagonist. Mycophiles gravitate toward one another, and at times we forget that fungi are often regarded with suspicion among the public in general, and Portland customs officers in particular. Having finally decided that neither I nor the mike-oh thingies posed a threat to the future of the United States of America, my interrogators abruptly expelled me into the arrivals hall.

The situation improved rapidly with Sallie's warm welcome and the midnight feast she magically procured from the boot of her car. Munching on her homemade treats, we headed north-east through the foggy night toward Mount Adams in the Gifford Pinchot National Forest to join the annual gathering of the Pacific Northwest Key Council. The council brings together the clever people who formulate diagnostic keys that help unlock the identities of Pacific Northwest fungi. They meet each year to share discoveries and search for fungi in the old-growth forests of the Cascade Range and beyond. Sallie crouched over the wheel, braking intermittently when a skunk or an elk loomed in the headlights, but undistractedly reeled off the names of the Key Council mycologists and their areas of interest. I recognized their names as the authors of the many fungus field guides and mycological texts on my bookshelf. I could hardly believe my luck in finding myself among such taxonomic expertise.

The following morning, I headed off with my new companions to our first field site, Dead Horse Meadow. I was pleased that it didn't live up to its name and instead produced an impressive suite of wonderfully unfamiliar mushrooms. Ducking through a blazing understory of vine maples, I saw that the forest floor was crammed with dense clusters of tiny fungi known as scarlet fairy helmets. Interspersed between their brilliant burnt-orange caps were the convoluted lobes of the hooded false morel. Nearby we found the widespread but uncommon blue chanterelle, and the unusually scaly vases of the woolly chanterelle. Scanning the undergrowth beneath the Sitka spruce, it was the lobster mushroom that had me stumped.

If I were asked to nominate the strangest fungus I have ever encountered, it would be difficult to isolate a single species from a long list of curious candidates. Fungi by nature are diverse and peculiar. However, on the shortlist would be the lobster mushroom. This fungus manifests as striking orange and oddly contorted goblets. How it got to look like that is a whole other remarkable story.

The name "lobster mushroom" is somewhat misleading. It isn't a species per se, but the merging of two. It results from a parasitic fungus, *Hypomyces lactifluorum*, infecting a mushroom. Here in the forest the parasite was infecting the stubby brittlegill, but it occasionally infects other brittlegills as well as milkcaps. The transformation that occurs during the parasitization process is radical. The parasite completely engulfs the host mushroom in a stroma—a crust of fungal tissue—distorting it and attiring it in a new brilliantly orange outfit. Once complete, the host mushroom is barely recognizable as its old self, with the parasite effectively availing itself of the host's entire anatomy.

Although the lobster mushroom retains the general shape and form of the stubby brittlegill, I was astonished by the makeover of its underside. Where one might expect to see the lovely radiating lamellae of the brittlegill, this surface was smooth. Perplexed, I checked another mushroom, then another. Some had a hint or trace of lamellae in the form of subtle blunt ridges. Examining one closely, I saw it was minutely pimpled by the parasite's tiny spore-packed perithecia embedded in its crust.

In the process of infection, the parasite apparently halts the development of the mushroom's lamellae, not only effectively castrating it, but taking up residence in the available space to manufacture its own spores. Some of the lobsters I examined had their stipes and lamellae engulfed by the parasite, but the rest of the mushroom appeared to be uninfected. It seems the infection process begins with the mushroom's fertile anatomy and then spreads further. To first take command of the host's reproductive potential by smothering its spore-producing surface seems a clever strategy. Little is known about the infection process and how each species responds and transforms, but as the takeover process progresses, the parasite's DNA increases while the host's DNA diminishes. I asked mycologist Fred Rhoades what he thought was going on. He suggested that there might be some parasitic hyphal connection through the mycelium. Fred's theory seems plausible, because when you find one infected mushroom, others nearby are similarly infected, with few, if any, escaping the parasite.

This enterprising fungus reminded me of cuckoos, known as brood parasites. These resourceful birds lay their eggs in the nests of other birds, abandoning their parental responsibilities and leaving the hard slog of chick-rearing to an unsuspecting foster parent. In doing so they save an awful lot of energy, allowing themselves more time for creating offspring than rearing them. Likewise, gall wasps manipulate plant cells to form nutrient-rich nurseries for their young in the form of swellings called galls. Creating a mushroom is a similarly large investment for a fungus. Working out a way to commandeer one that is ready-made makes good sense. There are countless similar examples across kingdoms, and some fungi have mastered a parasitic existence. Wherever there's a fungus, there's likely to be another trying to colonize it, be it parasite or saprotroph.

I scanned the forest floor to see that dozens of infected lobster mushrooms surrounded me. Dressed in their high-vis attire, they were easy to spot. Some were in shades of orange while others had become a blotchy wine-purple as they'd aged. Although the infected mushrooms retained their general form, they were oddly distorted and disfigured by the parasitization process. Why the mushroom contorts in the invasion process is not fully clear. I wondered if the contortions helped the parasite maximize the surface area over which it could release its spores. Cartoon images of the host mushroom madly thrashing around trying to rid itself of the wretched invader and contorting in the process came to mind, but all was calm on the forest floor.

The parasitization process brings about other changes. If you give lobster mushrooms and uninfected stubby brittle-gills a comparative squeeze, you notice that the host becomes firmer as the parasitization process progresses. Perhaps the firmer flesh, along with the greater surface area created by contortions, increase its resistance to deterioration. That is, the parasite to some extent might "mummify" the host to prolong the existence of its food supply. But what happens to the parasite once all the lobster mushrooms collapse, decompose, and disappear for the season? Where does it go? I wondered if the parasite is constantly present in the soil or tucked up among the mycelium of the host.

Some of the lobster mushroom's relatives within the genus *Hypomyces* form what are known as chlamydospores. These spores are thick-walled, often darkly pigmented, and rich in lipids (fats). They are effectively survival structures that enable the parasite to withstand adverse conditions or periods between seasons when the host fungus isn't producing mushrooms. Chlamydospores, however, have not been associated with the lobster mushroom. Yet the fungus must exist as propagules in some form in the soil, either as mycelium or spores, which are somehow alerted to the presence of a host. But it hides itself well, and to date it has never been found in environmental samples. Searches of genetic sequence databases, such as GenBank, reveal samples only from actual specimens, not from the environment. Additionally, it does not appear in fungal culture collections. This points to obligate parasitism—a parasite that depends entirely on its host for survival and reproduction. Mycologists have been pondering such parasitic conundrums for over a century, yet for now it seems the lobster mushroom is keeping the complexities of its lifestyle and its hideout to itself.

Although the lobster mushroom's complicated life history and perplexing form remain unresolved, its culinary value is well established. While the stubby brittlegill is not poisonous, its flavor is commonly considered bitter or nondescript. That is, it might be vaguely edible, but it's certainly not palatable. However, once the brittlegill is parasitized by *H. lactifluorum* and transformed into a lobster mushroom, its flavor improves. Foragers and chefs wax lyrical about its taste and texture. The visual appeal of it appearing like its namesake could be as much a flavor enhancer—at least for the adventurous—as the idea of eating something weird. Others might need more convincing.

Molecular biologist Genevieve Laperriere and colleagues documented the metabolic transformation and chemical changes that occur during the parasitization process. Their analyses showed radical chemical changes in lipid, terpenoid, and amino acid content in the host. Terpenoids are a large and diverse group of naturally occurring compounds found in most living organisms. They are well known for their role in plant defenses against stress and as signal molecules to attract pollinating insects. Terpenoids are also thought to chemically orchestrate interactions between fungi and other organisms, and their environments. Some fungi use terpenoids to lure spore-dispersing animals or to mediate communication with bacteria. Other fungi employ them as toxic weapons for capturing nematodes. Stubby brittlegills are thought to produce terpenoids partly for protection against infection. However, it seems they are a poor defense against *H. lactifluorum*: as the stubby brittlegill becomes infected, its terpenoids disappear. The parasite is an accomplished choreographer that controls the host's ability to resist it and deactivates its reproductive potential. While these chemical transformations spell disaster for the stubby brittlegill, they provide a coincidental culinary benefit to humans. Terpenoids are also thought to be responsible for the bitterness of the stubby brittlegill and hence their elimination improves their flavor.

Hypomyces lactifluorum is an unusual example of a *Hypomyces* that increases the palatability of its host to humans. Other *Hypomyces* (such as *H. perniciosus*) infect fungi including the commercially produced button mushroom, causing wet bubble disease. The combination of mushroom malformation; a fuzz of mycelial growth; and an unsightly, malodorous brown liquid renders them unmarketable. The disease can cause significant yield and economic loss for mushroom growers. From the parasite's point of view, it's just doing its job, stopping any one species from dominating, even if it happens to be someone's monocultural mushroom crop.

On my visit to the to the Pacific Northwest, I was curious to find out how the lobster mushroom fitted into the region's famous trade in wild mushrooms. Itinerant mushroom pickers scour the forests for sought-after species that they collect and sell to buyers. The buyers sort, grade, and clean the mushrooms for wholesalers, from where they quickly end up at farmers' markets, restaurants, and gourmet delicatessens. Commercial mushroom harvesting grew alongside changes in the forestry industry in the 1970s and 1980s. Many of the original pickers were ex-forest workers, but today many migrants are taking up the challenge.

After an enlightening few days with Sallie and the Key Council mycologists, I said goodbye and continued my journey further north in the Cascade Range to Mount Rainier, where I was keen to explore the fungi of its subalpine meadows. On the way I pulled over at a "Mushroom Buyer" sign, then tentatively poked my head through the door of a large metal shed. From floor to ceiling it was crammed with crates full of mushrooms. Buckets loaded to the brim with more were lined up on the floor. A head popped out from a behind a crate and a tiny woman in a stripy poncho beckoned me inside.

Isabella Romero was grading king boletes. Her hands flicked around a mushroom as she whittled off the damaged bits with a knife. I was obviously not a picker coming to sell my foraged mushrooms, nor did I look like an inspector. I was both amazed and grateful that she made time to talk to me, given I was just a curious blow-in. Isabella had been working since long before sunrise, cleaning, grading, slicing, and packing mushrooms. "Kings have to be processed within 24 hours of being picked," she explained. "Tomorrow, someone will eat them in a Seattle restaurant." The whole time she spoke, her hands didn't stop moving. I marveled that all her fingers appeared intact beneath the blur of the swiftly moving blade. Isabella then showed me her crates full of Pacific golden chanterelles and others with American matsutake, along with a box of cauliflower mushrooms, before taking me outside to see her mushroom dryers.

Out the back of the shed, Isabella's husband, Carmelo Martinez, stoked the wood fire of a mushroom dryer built on a flat-top trailer. Carmelo nodded a greeting and flashed a gap-toothed smile. Like Isabella, he looked exhausted, with dark rings under his eyes and blackened hands from handling mushrooms. Isabella pulled out trays of the dryer to show me their drying king bolete and lobster mushrooms. I asked Isabella about the lobsters. "In Oaxaca, we call them *oreja de puerco*. That's pig ear. Not lobsters," she laughed. "But they keep their color and firmness when you cook them." Isabella hails from the southern Mexican state, where her family's mushroom-collecting traditions go back generations. These ones were destined for the markets in Oregon and some were heading further south to Mexico. Isabella reached for a large jar of dried lobsters, unscrewed the lid and said, "Here, smell them!" The drying process had intensified their scent, and indeed they smelled of a distinct combination of forest and sea.

As we stood talking, a group of mostly Asian men appeared, dressed in camouflage pants. Each held buckets loaded with mushrooms. Carmelo nodded and went over to greet them. I watched curiously. The men bantered, squinting through cigarette smoke. Carmelo picked a large king bolete out of the bucket and gave it a gentle squeeze. He then shook his head and clucked his tongue. This met with protest from the men. Carmelo took out a knife and sliced through it. The men leaned in to inspect the dissected mushroom, checking for telltale traces of infection by maggots. The banter continued and more mushrooms were inspected, sniffed, and squeezed. Mushroom buyers grade each mushroom—the higher the grade, the higher the amount paid. Another group of men with mushroom-laden crates appeared behind us. Isabella looked at me, shrugged her shoulders, and continued sorting mushrooms.

I figured it was a good time to leave, thanked her, and ducked away, leaving the men to negotiate their deal. I've spent my life in the forest trying to understand the ecology of fungi. This cultural crossroad of forest and commerce was fascinating, but it felt like far more treacherous terrain.

7
THE MYCOPHAGISTS

I'm waiting for the sun to rise. Bands of fog settle in the forested folds of the Jura Mountains in Switzerland's far north-west, revealing themselves with the first light. In the deep limestone gorge below, the Franco-Swiss border traces the twisting course of the River Doubs. I pull on my coat, grab my camera bag and descend through the steep forest of spruce and beech, inhaling its heady scents. A heron observing my descent from the opposite riverbank stands in France. The trees around me are luxuriously padded with mossy upholstery, bulging as if stuffed into too-tight sweaters. They're like something from the mind of Dr. Seuss. I press against them and breathe in their moist vitality.

There is something captivating about the Doubs. It is tranquil yet vibrant as it slides by, bubbling and gurgling. Unlike its more highly regulated reaches, here there are eddying backwaters and riffles, vegetated islands and overhanging trees. They provide diverse niches for aquatic lives. The surrounding forest—protected by the Theusseret Forest Reserve—is similarly diverse, with some ancient woody elders that escaped the ax in former times. A range of tree species of different ages and lots of fallen wood make it rich terrain for fungi. And the fungi make it a good hunting ground for mycophagists—those who like to eat them.

The relationships between fungi and plants are becoming better known. Interactions between fungi and countless different animals have been less explored, but both fungus and creature can benefit from the deal. Fungi provide shelter and food for animals, with many fungi having higher concentrations of phosphorus, nitrogen, and vitamins than in the surrounding leaf litter. By consuming fungi and excreting them, animals disperse spores and therefore help sustain and perpetuate fungal diversity. This is especially important in disturbed systems, such as after logging or fire, when fungus communities are lost or dwindling.

Human foragers know that if they are to enjoy their bounty they need to find fungi that are uninhabited, although that's easier said than done. The slice of knife can expose a maggoty menagerie tucked up inside the inner depths of an otherwise perfect-looking Périgord truffle. Few fungi are unoccupied. Morels are especially popular hideouts. The cavalcade of wiggling, slinking, and squirming creatures exiting an upturned morel suggests the diversity of inhabitable crannies and crevices provided by their convoluted caps.

Mammals also actively seek fungi. Once, while foraging further south in the Jura Mountains, I got caught in a steadfast standoff between two mammalian mycophagists. My foraging companion Barbara Thüler and I came across a handsome, hefty porcino accompanied by an industrious mouse eagerly nibbling its way across its cap. Barbara took out her knife to remove the mushroom as the mouse dashed for cover. I felt a sudden sense of alarm. "But wait, you can't take its breakfast!" I protested. Barbara paused, then turned and looked at me, incredulous. I knew how much she adored eating porcino, and this was an especially impressive specimen. I hesitated. "Okay, then what about taking half? Half for you and half for the mouse?" I reasoned. Barbara could hardly believe that I was depriving her and defending a rodent but, with an obliging sigh of surrender, she obediently cut the mushroom precisely in half. The mouse was now nowhere to be seen, but I felt a satisfying if sheepish sense of relief. Despite this incident happening over a decade ago, Barbara has never let me live it down. And never will. Whenever I've

The glowing ghost fungus (*Omphalotus nidiformis*) is conspicuous for its bioluminescence.

Moss and fungi including lichens colonize an old log.

Many Māori people of New Zealand know the
coral tooth fungus (*Hericium coralloides*) as *pekepekekiore*.

The bloodred ruby bonnet (*Cruentomycena viscidocruenta*)
grows in moist forests in Australia.

The concentric-zoned sporing bodies of the polypore known
as turkey tail or rainbow fungus (*Trametes versicolor*)

Smooth cage fungus (*Ileodictyon gracile*) attracts humans with its stunning lattice-like structure and insects with its foul odor.

The long, ropy mycelium strands of the honey fungus (*Armillaria mellea*) allow long-distance nutrient and water transport.

These psilocybin mushrooms or magic mushrooms helped popularize fungi.

Angel wings (*Pleurocybella porrigens*) emerge from the side of an old hemlock log.

The scarlet bracket (*Pycnoporus coccineus*) colonizes a burnt log.

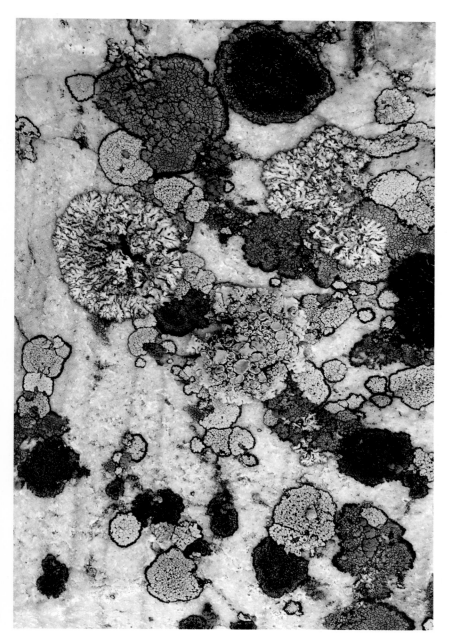

A palette of lichens claim territory on bare rock.

Mycelium consumes a eucalypt leaf.

A felty, fibrous cap typical of the genus *Inocybe*, known colloquially as fibrecaps

Tasty sliced lobster mushrooms (*Hypomyces lactifluorum*) drying in preparation for export

Isabella Romero prepares a king bolete (*Boletus edulis*) for market.

Traditionally, some have associated mushrooms growing in rings or arcs—such as these cloud funnels (*Clitocybe nebularis*)—with the workings of witches.

found a porcino since, she looks at me with feigned disdain and asks, "Don't you think it's best you only take half, in case there's a hungry mouse around?"

Spineless Alliances

Zigzagging through ferns and horsetails along the Doubs, the first fungus I encounter is a common puffball. It may be called common, but it is not ordinary. Vernacular names that reduce things to "common" seem unimaginative, especially for stunning fungi with distinctive features such as those of the common puffball. Its species name, *perlatum*, means "pearl," and indeed its entire surface is studded with tiny gems, although if you look closely, they are pyramidal spines. It resembles a miniature mace. As the puffball ages, the spines fall off, leaving a slightly asymmetrical imprint, reminding me a little of a sand dollar.

This particular common puffball is not only losing its spines, but it is punctured with holes. Dropping to my knees, I peer inside. The holes lead to chambers that house multiple slugs. It seems like a terrific opportunity to capture some video footage of slug lives within, so I set up my tripod and camera on time lapse. There is, however, a slight problem. The slugs don't appear to be doing much. In fact, they're not doing anything. All seem to be sleeping. I'm not in any hurry, so I settle in with my camera. An hour passes. The slugs sleep. I watch a woodpecker doing a little home renovation in a nearby beech tree and the fastidious activity of ants repositioning spruce needles in their mound. The river meanders by, and in a flash of blue a kingfisher skims the surface. The slugs continue to sleep. Hoping for a little action, I tentatively tap on the outer wall of the puffball with a tiny twig. No response.

Feeling the cold settle into my bones, I reset the camera and wander off through the forest, hoping, as always, that I will not be wandering around at midnight trying to find it again. I venture further down into a gully where frogs croak in unison, reveling in the dampness. After another hour, I manage to relocate my camera and peer through the holes in the puffball. It is entirely empty! The slugs have gone. It seems they were not so sluggish after all. I excitedly preview the footage to witness them exiting their holes. It was not only the slugs on the move. Other slugs had slid in and out of the holes, trailed by a tiny hairy snail. Flying squadrons of winged critters, mostly flies of various kinds, also visited. A fat red mite attempted an ungainly circumnavigation. A ladybird made a dramatic crash-landing before bouncing out of the frame. Perhaps the slugs were awoken by the racket on the roof of their chamber and decided to move on.

I am spellbound by the drama and exchanges in the tiny microcosm of a puffball that took place in my absence— dozens of creatures and liaisons, playing out on the puffball stage. I replay the footage over and over, spotting new characters with each viewing and marveling at the enterprise deep within the catacombs of a single puffball.

If you sit with a mushroom for an hour, you're likely to lose count of its many spineless guests. Or if you make a time-lapse film, you can view their antics in the wriggling and hopping, sliding and scurrying of various processions. When it comes to finding shelter, some people prefer five-star while others are cool with camping. Invertebrates also have preferences for the different fungi they inhabit, some are choosier than others. Slugs are partial to puffballs but especially favor brittlegills. The conspicuous holes and dots of white on the surface of brittlegill caps are an obvious clue. Slugs feed with their tongue-like radulas, which are equipped with thousands of toothy denticles. Radulas work rasping wonders. The dots on brittlegill caps are areas where the mushroom's surface layer has been grazed off, exposing the white tissue below. It can be tricky to find a brittlegill without these dots—they're a noteworthy feature of the genus. Slugs also like milkcaps, squeezing themselves into their often hollow stipes. The abundant squirming of maggots in brittlegills and milkcaps suggests they are also a favorite nursery of flies.

Just to the north of Hotel Puffball I spy a rooting shank. These elegant mushrooms stand tall on long slender stipes, making it easy to see beneath their caps. Their lamellae are a chosen haunt of springtails, gregarious creatures that graze on spores. I watch them crawling around between lamellae, and every now and then one senses my presence. Perhaps mistaking me for a predator, it rapidly propels itself out of the mushroom with the aid of a spring-like abdominal lever called a furcula. It seems like an exhilarating way to get around.

Among the assortment of other spineless life, little red mites amble around everywhere, probably seeking spores or other unsuspecting invertebrates to eat. Mites and other creatures with eight legs (arachnids) have a different gait from those with six, making particularly odd maneuvers, like some kind of wacky line dance. It's a helpful clue for recognizing them if you don't have time for leg-counting. Mushrooms are also often woven with the webs of spiders and other secretive spinners. As predators, spiders are unlikely to be eating mushrooms, but their presence suggests they are ready to pounce on those organisms that do. And what better way to keep dry than to do so under the shelter of a ready-made umbrella? Other creatures also seek mushroom shelter from predators, rain, or desiccation.

Spiders snare prey in webs but other trappers catch spores in nets. Mycologist Tom May recalled rolling over a log in his suburban Melbourne backyard and noticing the web-like nets beneath a fungus. "It's not a spun-silk net, but a slime net. The larva of what's probably a fungus gnat inhabits a slime tube within the slime net," Tom explained. He examined the net and noticed that it was covered in fungus spores. "Although the fly larvae are predatory, they could also be using the net to collect spores." Being a mycologist, Tom's focus is on fungi, but he was clearly enthralled by the interplay between fungi and invertebrates. "When you look closely, whole other worlds within worlds open up, and we know so little about what's really going on. I usually think of these interactions between organisms as happening in remote forests like the Tarkine in Tasmania, but here they are in my own backyard!"

Not all are as enamored of gnats and co. as Tom. Some human foragers who eat the bolete known as a slippery jack have experienced an unpalatable crunchy surprise in the form of millipedes. Millipedes are partial to slippery jacks and seem undeterred by the glutinous covering typical of this genus. Other invertebrates also seek the shelter of slippery jacks. Fungi with pores provide ready-made cells or chambers of various shapes and sizes that make ideal havens to raise maggoty progeny into adulthood. Flies are huge fans of fungi, and, despite being maligned, they are meticulous creatures. They devote a good deal of time to grooming, systematically preening each bristly appendage, and often do so on mushrooms (and pretty much any other available surface).

Big red-eyed *Tapeigaster* flies especially favor fungi. I try to surreptitiously film them, but they are always acutely aware of my presence. They stand their fungal ground as I nudge my camera closer. If I reorient to find a better angle, the plucky fly will follow suit, repositioning like a boxer in a ring, tilting its head to face me. It's humbling to be intimidated by a fly. A *Tapeigaster* fly on the upper surface of a mushroom is usually a territory-guarding male. If another male arrives, a scuffle usually ensues, until the challenger retreats or the resident male is overthrown. Sometimes females can be seen on the underside, inspecting lamellae for places to deposit their eggs, which they insert into these soft and supportive fungal nurseries.

Field naturalist John Walter has also spent time observing these flies and notes that size does matter, with male flies preferring fungi that produce larger mushrooms, like the Australian parasol or skirt webcap. It makes sense for a male fly to commandeer more territory for his future family to occupy. "I suspect it may be partly about having a larger stage on which to strut and attract a female, and secondly, about ensuring a plentiful food supply for the larvae," John suggests.

I wondered how invertebrates make choices about which fungi to eat or occupy. Are they attracted to the odor or texture of a particular fungus, or its palatability or nutritional value? Or is it simply about the availability of real estate? When not dispersing themselves via wind or water, I imagine the home range of your average springtail is limited to nearby mushrooms. Perhaps some invertebrates are content with whatever mushroom pops up in their patch. Other invertebrates might seek out a particular favorite. There could even be those who prefer a mushroom with a view, which could be advantageous if there's a predator lurking nearby.

Noticing the liveliness of the forest floor means switching scales and zooming in to details. Mycophile Sequoia Lewien has told me she never goes foraging without her camera. Finding a mushroom, squatting down, and seeking different angles to photograph it allows her time to finely tune in to her immediate surroundings. "So many spider webs are strung between mushrooms. And what a great place to live for a spider, with so many insects!" she observes. It was because of her own stillness while photographing that Sequoia became acutely aware of all the movement on the forest floor. "There's just so much going on that you couldn't possibly notice at walking pace."

Once a mushroom has released its spores, its "purpose" is fulfilled. As it starts to decompose, volatile compounds, often with strong odors, are released. It is during this decomposition process that many invertebrates are drawn to fungi—and humans are repelled by them. Other fungi don't wait until this stage to stink and actively deploy putrid odors to lure flying vectors. Stinkhorns, when at their slimy spore-producing phase, are the most irresistible of all. Among the hordes of flies drawn to stinkhorns, I've spotted beetles, wasps, and bees. It's hard to know whether they are attracted to the fetid fungus or just the excitement of the party. They may well be incidental visitors, blown off course en route to a pollination destination. When stinkhorns appear en masse, their foul stench can ward off *Homo sapiens* long before they are within sight.

More agreeably, aniseed funnels release a sweetly fragrant scent that permeates the forest, drawing me to them, and conjuring images of an illicit invertebrate banquet furnished with ouzo and licorice. This fungus is unmistakable for its scent but also for its color. Somewhere between turquoise, teal, and duck-egg blue, these tones are unusual in fungi. Despite my imaginings, the spineless seem indifferent to its evocative aroma and striking hues, and are firmly focused on the stinkhorns.

While many fungi inhabit the forest floor, I remind myself to look up for those with higher ambitions. Bracket fungi or polypores appear like hooves, pancakes, shelves or brackets on the sides of trees. They're often tough and woody, offering longer-term residences to invertebrates than the more ephemeral mushrooms. Researchers in Poland's Białowieża National Park studying bracket fungi found that the more advanced the state of decay, the greater the diversity of inhabitant invertebrates. They consider decaying bracket fungi as biological hotspots, as they support high numbers of species within small volumes. Invertebrates such as some scuttle flies are largely found only in these fungal habitats. Similar trends occur in decaying wood. The diversity of both invertebrates and fungi increases as their habitats age; generally speaking, the older the forest the greater the diversity of species. As invertebrates tunnel through wood, they bite into it with their mouthparts, breaking it down into smaller pieces. This alters the wood's structure, topography, and chemistry, creating diverse microhabitats for a range of saprotrophic fungi.

Invertebrates are attracted not only to mushrooms and other types of sporing bodies, but also to mycelia. A population explosion of springtails can make a meal of a fungus mycelium. Because invertebrates favor some fungi over others, their movements and masticating affect the composition of fungus species in a habitat. Invertebrates can also influence spatial changes in a mycelium, the enzymes it produces, and how nutrients and water move in and out of it. Mycologist Lynne Boddy and colleagues conducted grazing experiments using springtails with various fungi. They've shown how fungi can respond to being eaten by modifying the shape and form of their mycelia, perhaps to make themselves less palatable or to somehow trick the springtails. Astonishingly, the part of the mycelium that changed was not the bit being eaten but the more remote sections. This suggests that a fungus mycelium can relay information about being eaten to other parts of its mycelial network, probably via chemical or electrical signaling. Similar warning responses are known to occur in plants when they're attacked by insects. The mycologists' grazing trials were conducted in the laboratory. How these relationships play out in the field within the more complex dynamics of forests is less well known.

Autobahns of ants traverse the leaf litter. An envoy of defectors tracks up a leg of my tripod. Along with other invertebrates that transport soil between surface and subsurface, ants are likely to disperse fungus spores that attach to their bodies and appendages. They consume further spores with food and expel them elsewhere. Other ants, such as the famous leaf-cutter ants of the tropical forests of Central and South America, have formed remarkable unions with fungi. These ants cultivate gardens of fungi in vast interconnected underground galleries. The ants collect leaf and other vegetation fragments, then feed them to the fungi. The fungi digest the vegetation and convert it into nutritious hyphae, which the ants then feed to their larvae. Relationships are complex, and a third player—an antimicrobial bacterium that the ants produce on their bodies—serves as a secret weapon against parasites that threaten to invade the fungi.

On the forest floor, the light begins to fade through my camera viewfinder. Time stands still in the forest and yet it has somehow whizzed past. When operating at the scale of the spineless, little ground is covered. I wondered how fungi deal with issues of scale and spore dispersal, especially when slugs sleep on the job. One group of fungi seems to have figured out that highly sprung legs make excellent vectors, and so they make use of mammals to disperse spores over long distances. In Australia, on the other side of the world from the Jura Mountains, there are more than 350 mammal species, some of which are expert at sniffing out truffles and sprinkling their spores.

Mycophagous Mammals

A pointy nose pushed through a pajama top, followed by an oversize foot. Both were attached to an eastern bettong that blinked at me, seemingly in disapproval, then retreated. Also known as rat kangaroos, bettongs are small, rabbit-sized marsupials endemic to Australia. They belong to the same family as potoroos. Once common in Australia, all bettong species are now under threat. Habitat loss, along with foxes, feral cats, and the wrath of some farmers who regarded them as potato-thieving rodents, drove the eastern bettong to extinction on the Australian mainland. Now they're back. And not always in pajamas.

I met ecologist Millie Sutherland Saines in the old sheep-shearing shed at the Mulligans Flat Woodland Sanctuary on the edge of Canberra, Australia's capital city. Millie had been raising the abandoned baby bettong known as Nancy. "I named her in honor of the herb, early Nancy—its tubers are a favorite food of bettongs," she explained. Millie told me how bettongs share an odd trait with other macropods, the group of marsupials that includes kangaroos and wallabies—they sometimes toss their joeys from their pouches, often in response to being disturbed. Having been ejected, Nancy was too young to fend for herself, but she seemed snugly settled into her pajamas and was eagerly devouring her specially formulated baby macropod milk. Soon, however, she'd be on the hunt not just for her tuber namesake but for tasty truffles.

The Mulligans Flat Woodland Sanctuary is Australia's largest publicly owned box gum grassy woodland. These woodlands were once a familiar part of the Australian landscape, with their widely spaced trees spreading over a grassy understory. However, intensive clearing for agriculture; and subsequent cropping, chemical use, and overgrazing has reduced them to less than 15 percent of their original area. Established in 1995, the 3,000-acre woodland is classified as critically endangered under national, state, and territory legislation. A predator-proof fence ensures its inhabitants can live free from the threat of predation by feral animals. The bettongs were reintroduced in 2012 and have since tripled in numbers. Other mammals are being considered for reintroduction. As part of restoring the sanctuary's woodland, Millie's colleagues at the Australian National University brought in large old logs and other organic matter. They provide habitat not only for animals but for fungi.

For a fungus, relying on wind or invertebrates to spread spores has its limitations. Both can be unreliable and neither carry spores very far, usually only a few yards at most. Truffles have a different approach to dispersing their spores. They retain them in their bodies until a mammal decides to make a meal of them. Taking advantage of a bettong, potoroo, wallaby, or other digging mammal is a better bet for spreading spores further. Researchers found that wallabies disperse truffle spores hundreds of yards and occasionally almost a mile. The spores of many truffles remain viable after traveling though mammalian digestive tracts, and, if they are deposited in a suitable location, will germinate and form mycorrhizas with plants.

Some truffles are thought to rely on both mammalian and invertebrate vectors. In France, mycologist Marc-André Selosse and colleagues revealed that slugs that ingest the summer truffle are more effective at degrading spore walls through enzyme attack than are some mammalian vectors. This improves the ability of the spores to germinate, and increases the colonization of tree roots by the fungus.

Australia is thought to have many more truffle species than Europe. It also has higher ratio of hypogeous (belowground) to epigeous (aboveground) species. The term hypogeous fungi refers to a range of fungi that produce sporing bodies underground. These include true truffles, truffle-like fungi, and false-truffle fungi. Sometimes these terms get mixed up or are used interchangeably, and often all are referred to as truffles—a term generally reserved for the famed gourmet truffles of the genus *Tuber*. They are also known as "sequestrate fungi," because their spores are sequestered and never released to the wind. Terms such as secotioid, gasteroid, and gastioid are used to describe different evolutionary sequestrate phases. The vocabulary can quickly become confusing. Without wanting to dumb down the complex lives of these funky underground lumps, from here on I'll simplify things by using the word "truffle" to collectively refer to the various types of hypogeous fungi.

Mycologists hypothesize that the production of underground sporing bodies is an evolutionary adaptation to climatic extremes. Not only is wind unreliable for spore distribution, but when it is combined with sun and the vagaries of weather, mushrooms can suffer a beating, withering and collapsing before their spores are mature and released. Living underground is a good strategy for keeping cool and moist. Many truffles are thought to have evolved from mushroom ancestors. That is, at an earlier time in their evolutionary history they had sporing bodies of the cap-and-stipe style. Over time, they've abandoned this anatomy, moving through various transitional stages to a life spent fully underground. Given there is no wind underground, the need for a stipe to push itself up through the soil and for lamellae to release spores to the wind became redundant. Hence, during the evolutionary transformation, the hymenium (spore-bearing surface) became enclosed, and the stipe was lost. Another group (which includes the true truffles, *Tuber* species) evolved from relatives of cup fungi.

Many of the truffles related to mushrooms have a form midway between a mushroom with cap and stipe, and an underground truffle in its lumpy ball. If you slice one of these in cross-section, you'll see the vestigial or "redundant" bits of its anatomy—the stipe and crumpled lamellae tucked up inside the ball. In addition to benefiting from the more consistent conditions of the subterrain, many truffles have a dense outer covering that provides an extra barrier against water loss. In dry woodlands or deserts, truffles, along with the odd lichen, bracket fungus, or puffball, are among the fungi more commonly found.

Some truffles are beautifully colored, but many appear as amorphous brownish lumps. They are not usually visually striking. However, when it comes to scent, some truffles are among the most breathtaking and intoxicating of all fungi. Truffles have evolved distinctive odors that are irresistible to countless mammals. These odors intensify with maturity and are extraordinarily diverse to a human nose—and probably more so to a bettong snout. Some are relatively mild and earthy. Others are fruity or garlicky. There are those that are sulfurous, sweaty or rancid. A great variety smell like cheeses, from roquefort to taleggio and gruyère, and many have a confounding cocktail of odors that are hard to distinguish or describe. Odors can also change at different stages of maturity. The odor of a truffle is perceived differently depending on who is doing the smelling. Experienced sniffers can separate out different "notes" and have the olfactory vocabulary to describe them. Pregnant women are often highly sensitized to certain scents and odors, while people with some illnesses may have a compromised sense of smell and find it difficult to detect them.

Some humans revere the European truffles of the genus *Tuber*, especially the famed Périgord truffle. Other mammals exploit hundreds of different truffles. More than 30 species of Australian mammals eat fungi, relying on them as food to varying degrees. Fungi typically have higher concentrations of nitrogen, phosphorus, and potassium, offering more nutrition than many plants. Obligate mycophagists actively seek and derive most of their nutrition from fungi, usually truffles. Preferential mycophagists also mostly eat fungi, but will settle for other food sources when they can't find themselves a fungal feed. Members of the rat kangaroo family (Potoroidae), like Millie's bettong Nancy, may be obligate or preferential mycophagists but are considered truffle specialists. Strong forelegs and claws adapted for digging through dry and hardened soils, as well as their long, pointy snouts, equip them well for truffle hunting. Bettongs and potoroos are thought to eat the greatest range of species, with up to 90 percent of their diets being truffles. Forty-six fungus taxa, almost two-thirds of them truffles, have been found in bettong feces.

Many other mammals are mycophagists too. Opportunistic or generalist mycophagists will eat truffles along with other fungi if they can't find their preferred foods. Among them are native bush rats; and marsupials such as pademelons, wallabies, bandicoots, and possums. Quolls (marsupial cats) and various birds are also thought to accidentally consume fungi while searching for other foods.

Mammals distribute spores, and their digging in search of truffles and tubers turns and renews soils. Ecologists dub them "ecosystem engineers." Ecosystem engineers directly or indirectly influence the availability of resources for other species by maintaining, modifying, or creating habitat. Wander around in an area where they're active and you are likely to see dozens of holes—or, as my friend Marina Lewis refers to them, "snoles," for snout holes. As with the work of spineless soil movers, these mammalian mix-masters help incorporate surface litter into the subsurface, increasing nutrient cycling and water retention. This in turn provides a source of food and habitat for organisms under the ground. It is all part of the process of building up biological activity and the organic component of soils. This collectively increases the vigor and resilience of plants, and improves overall ecosystem function. The movement of soil and increased heterogeneity creates new substrates, microclimates and niches for fungi, and redistributes fungus spores.

Digging mammals exist around the world, but are especially important in arid environments. As an adaptation to preventing water loss, Australian plants such as eucalypts have high levels of resins, waxes, and oils. When these are leached from decomposing organic matter, they can make soils water-repellent. Digging alters the structural and chemical properties of the soil, improving its aeration and water filtration. With more water entering the soil there is less run-off, thus reducing erosion. Plants in turn benefit from increased water availability and can more easily absorb nutrients. Digging mammals also affect seed dispersal and seedling success. Seedlings that grow in soil where there has been intensive digging activity do better than those that grow in less intensely dug soils.

Australia has a diverse range of digging mammals, but around half are threatened with extinction, and several have already succumbed. The fragmenting of landscapes and consequent disconnecting of mammals' habitats have devastated populations. Today the range of most mammals is radically reduced. When digging mammals are lost, other species disappear too. If a hypogeous fungus has no other means of spore dispersal than a digging mammal, and that mammal becomes extinct, the danger is that the fungus may also be lost forever. What happened to the truffles that relied on the extinct mammals as vectors? When an ecosystem is disturbed and the composition of species changes, so do the roles of and interactions between species. Mycologists are examining whether the loss of some mycophagous mammals is compensated for by others. This is known as functional redundancy.

Ecologist Susan Nuske has sequenced the DNA of truffle spores in scats to determine whether specialist mycophagists, such as the endangered northern bettongs, contribute disproportionately more to truffle distribution than generalist mycophagists (such as bandicoots, rats, and mice). She wanted to know whether a diverse group of generalist mycophagists would compensate for the loss of a specialist mycophagist. Would the same diversity of truffles be dispersed by the generalists in areas where the specialists were lost? Susan told me that northern bettongs not only ate more types of fungi, but five times as many unique types of truffles (those not eaten by generalists). In short, this means that if specialist mycophagists are lost from an ecosystem, the generalists are unlikely to do the job for them. Truffle species could be disappearing beneath the soil before their existence is even known.

As awareness of fungi grows, so does the popularity of foraging by human mycophagists. However, endlessly overflowing baskets of foraged mushrooms is becoming a thing of the past as thinking changes about foraging. Given mushrooms are made of chitin, a tough structural compound that is difficult for humans to digest, eating a few edible mushrooms will leave you feeling better than eating many. Selecting just a handful of perfect specimens also reduces the chance of consuming bacteria and invertebrates that inhabit mushrooms as they age. It also leaves a supply for other hungry mammals. Mindful foragers encourage an ethic of care when it comes to fungi and forests, and are looking to Indigenous wisdom for a more sustainable approach.

Mindful Mushrooming

It was in the Cascade Range of America's Pacific Northwest that I chanced upon an elder with a worthy message. I met Esther Win Bo and her mother Su Aung serendipitously at Mount Rainier. It was Su's 96th birthday. Esther had offered to cook her mohinga, a traditional fish soup from their homeland of Myanmar, but Su had wanted chanterelles. I was lying in the undergrowth filming fungi when a pair of boots walked into the frame. Esther, searching for chanterelles, nearly stepped on me, giving herself a terrible fright. Su almost fell over laughing. We ended up spending the rest of the afternoon together, and I was delighted to accept their spontaneous invitation to join them for the birthday dinner.

Fungi take time to get to know and, in nine decades of foraging, Su had come to know them well. Her family had foraged for edible and medicinal mushrooms for centuries. She could still recall her first mushroom-foraging trip, with her grandmother in Myanmar, when she was five. "Oh, she made me get up very early—it was dark," Su remembered. "We walked down the valley and into the forest. I found the first mushroom and my grandmother made me smell it. Dirt went up my nose, and I sneezed and dropped it!" Su laughed and laughed at her recollection until her glasses fogged up and slid off her nose.

She went on to describe the many types of mushrooms she had learned to identify as a child. The fungi of Myanmar are not well documented, but with the country's range of habitats and microclimates they are likely to be diverse. With few, if any, professional mycologists in Myanmar, much of the knowledge of fungi is held by foragers like Su and other mushroom enthusiasts. Su was amazed by the sheer abundance of chanterelles and other edible fungi in the Pacific Northwest forests. While she appreciated their flourishing, she was disturbed by the quantities she saw harvested. "Have you seen them?" asked Su. "Buckets and buckets and buckets, all full of mushrooms. But one day, they'll be empty."

The room filled with the aroma of spices, frying garlic, and chilli. Esther brushed the dirt off the chanterelles with meticulous care. Vegetables were chopped and soon a delectable mushroom khao suey materialized from the steaming wok. Over dinner, Su explained how traditions of foraging in her homeland had been passed through generations. There were strict guidelines about where, when, and how fungi should be harvested. These weren't enforced by law or regulation. They were part of an unspoken system of principles and cardinal rules that her family applied to all foraging and hunting.

"It's all about respect," said Su. "For everything—for all lives, including crickets and mushrooms, even the ugly ones." She screwed up her face, presumably mimicking an "ugly one," but she still looked beautiful. "It begins with gratitude, responsibility." Su's philosophy relied on ecological understanding gleaned from generations of accumulated observations. She explained the importance of recognizing connections and always assuming that further alliances exist unseen. "And you must only take what can be eaten. No more. Nothing should ever be wasted," she said, pushing wisps of gray hair back off her forehead. "And always give thanks. And never"—she paused—"never cause harm." The philosophy of Su and her family brought together ecological knowledge and something more akin to spiritual beliefs. This held them accountable to both the ecological and the metaphysical. Such an approach is shared by those living in close association with nature across cultures, knowledge systems, and beliefs.

I thought about being on Country in Australia and how Su's words echoed those of Wiradjuri elder Uncle Neil Ingram. He speaks of his people as being one with the land. "To harm it in any way is to break trust with the land, and with yourself," Neil had told me. This way of thinking is shared by Indigenous Australians and those who see themselves as inseparable from nature. Biologist and North American Potawatomi Nation citizen Robin Wall Kimmerer echoes Su's and Neil's philosophies in her telling of the Honorable Harvest, which describes the canon of collective wisdom, principles, and practices that guide and govern human relationships with the natural world. When we regard other organisms as "kinfolk," says Kimmerer, "another set of harvesting regulations extends beyond bag limits and legal seasons." She describes how the dictum of only taking what you need can also leave a lot of wiggle room, as many people are unable to differentiate "want" from "need," and reminds us of the almost forgotten tenet of not only taking only what one needs, but only that which is given. In modern profit-driven economies, where cultures of exploitation are commonplace, how do we distinguish between that which is offered by the earth and what is not? Kimmerer suggests there is no one path or answer, but it might begin with tuning in a little more acutely to the natural world around us.

When I'd arrived in the United States on the same trip I met Su and Esther, mycologist Steve Trudell had supplied me with a kit of maps, fungus field guides, and "survival gear." Among them was a detailed chart titled "Washington State Personal Use Mushroom Harvesting Rules." Although I was more interested in photographing mushrooms than harvesting them, the chart was an eye-opener. I ran my eye down the grid of places, species, permit requirements, and limits to harvesting. In most of the listed locations, each forager can harvest up to five gallons of mushrooms. Harvesting was not usually limited by season. Permits mostly weren't required. The footnotes to the chart listed various requirements for how mushrooms should be cut; authorities to consult; state rules, laws, and regulations; and harvesting etiquette. It was a detailed account. Similar codes of conduct exist for fishing, hunting, and the taking of other "natural resources."

The harvesting rules are well intentioned. They are designed to minimize damage to the environment and ensure mushrooms are not depleted. "Bag limits" are likely to be based on some sort of calculation or model deduced from estimates of abundance. I thought about how hard it is to come up with a set of numbers when forests and the climate are in constant flux. Collecting accurate data about the abundance, distribution, and life histories of any species is difficult, but inordinately more so for organisms such as fungi that are less understood and difficult to quantify, and that have not been comprehensively surveyed. Decisions are made from the best available data. But are they adequate? Are they complex enough to detect less conspicuous patterns or changes in fungus populations?

I wondered how often the data plugged into such models are revised, and if the numbers are relevant today. The regulations arose in the first place because our species has largely become too disconnected from the natural world to gauge the impacts of our actions. Most of us have lost Su's philosophy of attentiveness and the Honorable Harvest that Robin Wall Kimmerer describes. Harvesting rules at least offer a guide to keep us on track.

The week before at the Key Council meeting I'd met mycologist Lorelei Norvell. Lorelei is warm and generous with her knowledge, which she imparts with a quirky humor. She works with a range of fungi but specializes in rootshanks, the fungi Lee Whitford had introduced me to on the Olympic Peninsula. In the 1980s, prompted by a perceived decline in edible mushrooms among members of the Oregon Mycological Society, Lorelei and her team set up the Oregon *Cantharellus* Study Project. It was one of the first studies to assess the effect of harvesting mushrooms on fungus productivity, or the capacity of fungi to produce mushrooms.

Back at the dinner table, Su expressed her concern about the quantities of mushrooms taken from the forest. "Su," I asked, "do you know about the Oregon *Cantharellus* Study Project?" Over the ten years of the project, which was run at Mount Hood further south in the Cascade Range, the researchers showed that harvesting chanterelles did not affect future production. These results were welcomed by mushroom harvesters.

However, there are those who are more cautious, and are concerned that the results could be taken out of context to justify exploitation of other fungus species. Like all research, the study has its limitations and, as Lorelei herself argues, the findings of a study of one mushroom species cannot be generalized to others. She laments that "the public tends to remember the quick (and often erroneous) take-home message and miss the subtleties." When it comes to understanding the complexities of fungi and forest ecosystems, nuance counts for a lot.

Su pondered my question about Lorelei's study, narrowed her eyes and nodded. "One species, one habitat, and the study began almost 40 years ago," she noted. "What is the story of the fungus today?" I thought about my own experiences of being in the forest 40 years ago and some of the changes I had witnessed since then. Some are part of naturally occurring cycles, others are the results of human-induced change. Today we have a more rapidly changing climate and different patterns of weather. In some places, mushroom harvesting has intensified as the demand for wild-foraged fungi increases. Globally, forests are being felled at the fastest rate in history, both "legally" and illegally. Between 2010 and 2020, the annual net loss of forests was more than 12 million acres (an area about the size of Slovakia, the Dominican Republic, or Costa Rica), and deforestation rates were substantially higher. This seemed like the bigger threat to fungi than foraging.

As we talked about Lorelei's decade-long study, Su leaned in, putting her hands flat on the table. "Ten years might seem long to us humans. But is it long for a fungus? Do we really understand the longer-term effects of taking so many mushrooms? Will these reveal themselves within our human timeframes?" I thought about her caution in assuming there are unseen relationships that we don't understand. Changes beneath the soil might not be immediately apparent above ground. We don't know how old and established the mycelia of those mushrooms were when Lorelei and her researchers began their study. While many fungi grow in undisturbed forests, others are stimulated by some level of disturbance. Some people, including Lorelei, suggest that harvesting could stimulate mushroom production, as long as the mycelium is not harmed. But how do we know if foraging harms mycelium or not?

A team of Swiss scientists has offered some insights. A long-term (29-year) study by mycologist Simon Egli and colleagues in Switzerland also found that harvesting of mushrooms did not affect future yields. While picking mushrooms does not compromise the fungus mycelium, the trampling that occurs as part of harvesting does cause some short-term damage to the mycelium, though the Swiss study found that the fungi continued to produce mushrooms once trampling stopped. However, such findings are not absolute, and depend on the degree of disturbance. How many trampling feet is too many? How might this change in wetter or drier years, or in countries like Australia that have few deciduous trees and less of a leaf-litter layer, and where until recently foraging has been relatively light? And, if mushroom production is reduced, how many spores does the fungus need to ensure its survival?

I marveled at Su's sharp mind and ability to meld her ancestral knowledge with Western science. Su conceded that the research was a good starting point in helping to understand potential effects of foraging, but suggested that if we all adopted a more sensitive approach, these studies might not be necessary. After dinner, we exchanged contact details, and Su offered me an advance invitation to help her celebrate her 97th birthday. We said goodbye and, as I drove back through the forest to my lodge, I thought about the foragers' chart and the well-intentioned efforts to monitor harvesting. How could we reliably separate the potential effects of foraging from those of other pressures, especially climate change? So many questions remain unanswered, but so few long-term data on fungus distribution exist with which to answer them. Without such data, it is difficult to assess the conservation status of fungus communities worldwide.

Treading Lightly

It is hard to remember a time when fungi and the forest might have been different. Changes are constantly occurring but are not always observed. Humans tend to operate in the moment. The current state of the forest and its fungus populations is considered "normal." The perception of first experience forms the baseline, an idea that underlies what scientists refer to as "shifting baseline syndrome": the gradual change in accepted norms for ecological conditions. It exposes our failure to recognize the scale of change, as successive generations view their first impression of an ecosystem as the "new normal."

I remember the infectious excitement of one group of Australian children on their first fungus foray in a local park. "Quick, come here, they're everywhere!" screamed a young boy in elation. In the garden bed he'd found impressive troops of redlead roundheads, an early colonizing species. But that was all. There was only one species. Before the local council had reduced every log, branch, and stick to woodchips, an established community of diverse fungi had flourished. While it was wonderful to witness the children's delight, they had been deprived of the great diversity of species that had thrived before the park's ecology had been homogenized. And the ecosystem had been deprived of this diversity too.

Change can be chronic yet not readily perceptible. We live so much more of life than we're inclined to remember. While the more attuned may be aware of small changes and historic differences, it's easy to be oblivious to accumulative effects over time. Consequently, when standards are lowered little by little, our expectations can change, and each successive generation is unaware of the environmental conditions of years past. Ecological decline in a particular place may be more extreme than scientific data indicate. Knowledge extinction occurs when we forget our own experiences. This magnifies when we fail to capture the experiences of previous generations. Su was concerned about the things that can slip through the numerical measuring of nature. Her barometer was based on her own experience and that of her forebears. It relied on knowledge being continuously passed down. So what happens if those things not only slip through the data, but through memory as well?

Su's accumulated knowledge carries not just the data and regulations of Western conservation, but obligation.

I recalled a conversation with Australian-based Italian forager Diego Bonetto. His foraging practice also taps into the generations-old knowledge of his predecessors. For Diego, learning to forage is "a process of initiation" that begins with the guidance of experienced elders. "In Italy, where I grew up, you wait for someone to take you foraging for mushrooms. Foraging for mushrooms only gets handed down to worthy people, who show respect for the place and the biome," he told me. "Identification of species is only one aspect of the harvest. There are other important aspects that need to be understood, like legalities, ethics, how to look after the resources, timing, and conditions."

Today there are mantras about "treading lightly" that remind us to treat the natural world gently. Codes of practice urge us to temper our taking. They are usually applied as an overlay to an activity, not an inseparable fundamental premise as to how that activity should be practiced. They are often lists of things we shouldn't do, rather than things we should. The underlying assumption is that we have little understanding of ecosystems or awareness of the broader implications of our influence.

Few Australian foragers come to foraging from traditions passed down through generations. Many people live their day-to-day lives disconnected from the places and species they forage. In our book *Wild Mushrooming*, mycologist Tom May and I explore the notions of ecological foraging and slow mushrooming. Neither concept is new. We simply reiterate old ideas and the same kinds of principles that Su describes. We introduce the terms as shorthand, to rekindle or repurpose past practices in the context of foraging in current-day Australia. We offer them as an entry point into bigger-picture thinking and discussions around foraging, ecology, and conservation.

These concepts aim to revive old ways and inspire an ecological and ethical awareness within the environmental and cultural landscapes of today. Our philosophy begins with understanding the ecological significance of fungi and their environments. This takes time. Accelerated approaches can only ever offer an abbreviated account, a truncated version of a kingdom of staggering profusion and complexity. Fungi, like people, cannot be known and understood straight away. They can be elusive and unpredictable.

Human mycophagists who look after both fungi and their environments can increase appreciation of fungus conservation, and consider the many other non-human mycophagists also on the hunt for a feed. Mindful mushrooming is about care, attentiveness, and developing deep knowledge. It's about revisiting traditions and keeping up with the latest research. Ultimately, it's about being aware how every choice we make reverberates in the world around us.

8

CONSERVING THE BIZARRE AND THE BEAUTIFUL

I doubt I'm the only person on the planet to be perplexed by some politicians. I was once asked by a member of the national Australian parliament to justify why conserving biodiversity mattered and was momentarily stumped. Where was I to begin? It's like trying to justify why we should conserve our arms or legs, our mountains or rivers, or the air we breathe. I'd wrongly assumed it was self-evident. It was as if he regarded flowers, frogs, fungi, and every other form of life as mere decorations in the landscape—or, more likely, as things that could obstruct the future developments he had in mind. He'd obviously made no connection between the perilous state of our planet and all the processes, functions, or services provided by the great sweep of life otherwise known as biodiversity. It was a bewildering experience.

If you are a numbat or a red-tailed phascogale, a night parrot or a windswept spider orchid, your future survival is at risk. However, at the very least, your plight has been recognized and there's possibly a dedicated team of advocates fighting for your survival. Beauty, cuteness, and charisma hold sway. Fungi, on the other hand, have fewer followers keeping track of their conservation needs, with fauna and flora being the focus of biodiversity conservation in Australia and elsewhere. But fungi are slowly infiltrating the conservation agenda.

In general, conservation issues for fungi are like those for fauna and flora. Habitat loss is a major concern. When habitats are disturbed, destroyed, or fragmented, populations of fungi and their spore-ferrying vectors are reduced and isolated. Hence, mycologists mostly opt for habitat protection as a vital umbrella approach to conserving fungi, especially given most are as yet unknown to science. Because fungi mostly live their lives out of sight in the subterrain, they are prone to unseen declines and extinctions, disappearing unnoticed.

Little is known about the distributions, life histories, and threats to most fungi. Some require very specific habitats, such as undisturbed native grasslands or old unburnt wood, so a single-species focus is needed alongside habitat protection. Given most conservation initiatives are based only on the needs of flora and fauna, it is not known if they effectively conserve the diversity of fungi. While protecting habitat is a worthy approach, it is only truly valid if it takes into account the needs of all the biodiversity it supports. To neglect fungi is to misunderstand how ecosystems function, and to overlook a fundamental element of biodiversity and conservation.

Discussions about conserving fungi usually focus on the tools and strategies for their protection, such as the IUCN Red List of Threatened Species, legislation, and policies. All of these form necessary frameworks for fungus conservation. Less often explored, however, are the tensions between fungi, aesthetics, and conservation, and how our perceptions as humans shape the ways we regard and value fungi. How might the appearance of a fungus affect whether we value it or not? How do we negotiate the intersections of aesthetics and conservation to rally public support *and* ensure best-practice conservation? Can art and aesthetics serve as tools for inspiring greater appreciation of fungi? Delving deep into the lives of fungi reveals some surprising examples of how their aesthetics both help and hinder their survival.

Dazzling Waxcaps

Every environmental cause needs its ambassadors and advocates. Flagship species are those species chosen to represent a group of organisms, an environment or an environmental issue. They're typically considered charismatic or beautiful, and are selected to attract broad interest, rouse public sympathy, or lure financial support, even though they may not be the most ecologically important species. Across the world, people are familiar with flagships such as pandas and polar bears, humpback whales and koalas. Such species are charismatic and usually inspire an emotional reaction in people. Given that many people respond to aesthetics and emotions before science, there is merit in choosing visually appealing species.

Waxcaps are fungal flagships with great aesthetic appeal. These striking fungi appear in a spectrum of colors from crimson, pink, and mauve to lime green and chrome yellow. Some transform from carmine red to jet black with the gentlest touch. With their striking hues and waxy caps, they are often so perfectly formed as to appear fabricated. In Europe, waxcaps are recognized indicators of undisturbed or "unimproved" grasslands—ironically, this refers to those free of fertilizers—although they are found in woodlands and forests as well. In the year 2000 and in a boon for Australian fungi, a conservation reserve was established to protect an endangered community of waxcaps, at Lane Cove Bushland Park in suburban Sydney. It's one of only about a dozen reserves worldwide specifically set up for the conservation of rare fungi. The listing of the reserve on the Register of National Estate set a conservation precedent by recognizing fungi in biodiversity legislation that had previously only included animals and plants. The park harbors not only rare and endangered waxcaps, but a previously undescribed species now known as *Hygrocybe lanecovensis*.

It's not easy for a fungus to acquire itself a conservation reserve. Determining the conservation status of a fungus is an involved and lengthy procedure. Its identity must be documented, and its distribution and rarity established. As fungi are seldom part of biodiversity surveys, it can be difficult to work out whether a species is truly rare or just under-surveyed. Although it's a rigorous and systematic scientific process, there's no denying that looks can help. The captivating images of waxcaps from Lane Cove Bushland Park may well have been a selling point, if not in the registration of the park then certainly in arousing public interest. If the waxcaps had been dull and less endearing, they might have attracted less attention.

A few years before the delightful Lane Cove waxcaps were assigned their reserve, another extremely rare fungus, known as teatree fingers, was found in Victoria. The brown lobes of this fungus, resembling clasping fingers, appear on the wood of a couple of tree genera and only those that have not been exposed to recent fire. Although it's been listed on the state's protective legislation, the *Flora and Fauna Guarantee Act 1988*, since 2004, the fungus received little attention until recently. Mycologist Sapphire McMullan-Fisher is working on a project to determine its distribution. She told me the welcome news that her surveys revealed teatree fingers to be more widely spread and in association with a greater range of tree hosts than previously thought. However, garnering public interest in a less spectacular species can be difficult, especially one that is thought to parasitize other fungi. The notion of protecting fungi is already challenging for some people, let alone protecting a fungal parasite.

Various conservation tools and strategies are used to protect species, and aesthetics and science both play a role. While aesthetics are important in selecting flagships, the nomination of a species for a conservation list relies on in-depth knowledge of its ecology, life history, and distribution. Many species have been assessed for their conservation status worldwide. Extinction risk is tracked through the IUCN Red List of Threatened Species, which has been compiled and published for almost 60 years. Today more than 147,500 species have found themselves on the list, 41,000 of which are threatened with extinction. Red Lists are also prepared at national and regional levels, and serve as an important conservation tool for prioritizing species for conservation. Yet they largely neglect fungi.

In 2013, in response to the near exclusion of fungi from the IUCN Red List, mycologists Greg Mueller, Anders Dahlberg, and Michael Krikorev established the Global Fungal Red List Initiative. The project coordinates assessment of the threat status of fungus species. Information on almost 1,900 species from 80 countries has been assembled, leading to formal assessment of around 500 species for the IUCN Red List. Although the listed species on the Global Fungal Red List Initiative have no legal standing, it theoretically brings them into the spotlight and onto international conservation agendas.

A quick whiz around the species included in the initiative reveals great variation in color and form, but few, if any, of the 28 species formally listed as Critically Endangered on the IUCN Red List would be considered charismatic. The lichen known as old birch spots, for example, hardly sounds enticing. It appears as a black-spotted grayish white crust, and it only grows on the bark of mature yellow birch at high elevations in the southern Appalachian Mountains of North America. The truffle-like fungus *Destuntzia rubra* appears as distorted pinkish lumps that emit a distinctively fishy odor. *Kavinia chacoserrana* manifests as a yellowish-cream spiny integument on decaying wood in the tabaquillo forests of Argentina. Only a dedicated mycophile would recognize it as a fungus at all.

Aesthetics and conservation have a complicated history, but the aesthetics of individual species is probably less of a hurdle to fungus conservation than the lack of attention to the entire Kingdom Fungi over the last two centuries. Red Lists and flagship species have increased public awareness of endangered species, and fungi are finding their way into these and further conservation initiatives. Growing public interest in fungi, the availability of cameras in mobile phones and the swift dissemination of images via social media and other online platforms have helped bring fungi into the mainstream. Fungi are not only being noticed, but positively represented.

Seeking Fungal Beauty

While aesthetics play little part in official Red Lists, there is no doubt that an attractive fungus can capture the imagination of the public, especially that of artists. Humans desire beauty. We admire stunning floral displays, and there are long histories of collecting beautiful butterflies and birds, often to the point of their demise. Displays of beautiful fungi offer a tantalizing new visual delight for many people, even if it is an unfamiliar or obscure beauty. Over the years of running fungus forays and workshops, I've observed how people are more likely to respond to beautiful, colorful, or eccentric mushrooms than the infamous LBMs— little brown mushrooms.

In the fall of 2021, I met with a group of fungus enthusiasts from the World Wild Fund for Nature (WWF) who were joining me on a foray. We wandered along the River Suze that carves through the Dovecote Gorge in the southernmost fold of Switzerland's Jura Mountains. Beech trees clung to the steep cliff sides, shedding their fall foliage into the swirling waters below.

The forayers had come along for different reasons. Some were recording the species they observed and keeping lists, consulting field guides or apps. Others were scouting specifically for edible species. Many were photographing. I watched them as they searched for fungi, and was curious about those that gave them pause, or that caused them to suddenly drop to the ground or to exclaim in excitement. I had a strong sense that for many of them, their fascination or interest was aesthetic. Some simply wanted to be around beautiful things. Beautiful fungi. While there were those compiling visual inventories by photographing every fungus they saw, others were on a mission to capture the most alluring fungus they could find. I recognized how the quest for beautiful fungi involves both science and aesthetics, and each can increase appreciation of fungi and the natural world.

As we headed deeper into the gorge, I paused by Bea Harris, one of the older participants in the group. She was adjusting her tripod to photograph a cluster of verdigris agarics among the moss on a rotting log. Squatting down beside her, I peered at the vivid turquoise mushrooms, their caps delicately sprinkled with white veil remnants. I've seen this fungus many times but am always struck by its color. "Oh wow, you've found *Stropharia aeruginosa*, or the verdigris agaric," I told her. "What a perfect bunch!" Bea looked at me and smiled, then gently told me, "It could be called Fred Astaire for all I care." I took her point. She simply wanted to delve into its beauty, to inhabit the wonder of her discovery, not have it stamped with a name. I thought of a similar experience of a dawn forest walk where my companion felt the need to name and explain each bird species we heard. But his unrelenting commentary spoiled my experience of reveling in the sublime music of their calls. Despite our curiosity to understand the fungi we find, not all of those admiring fungi for their beauty want the experience explained away by science. For some, beauty itself is enough to make them meaningful.

Fungi have largely been examined through the lens of science. However, art and mycology have long influenced each other, and the natural history of fungi is deeply rooted in aesthetics. Art helps illustrate the abstractions of scientific data, but it goes beyond that. Artists reveal new perspectives on and dimensions of fungi that broaden our perceptions and understanding, but they also critique the ways that scientific and artistic knowledge have been separated. Both science and art are profoundly creative processes. Each arises from curiosity, inspiration, and imagination, and relies on astute observation and interpretation. Mycologists and artists often draw inspiration from similar places, and can enrich each other's thinking.

Many mycologists were, and are, also artists. The captivating mycological illustrations of French mycologist, physician, and artist Charles Tulasne (1816–84), especially those in the three-volume *Selecta Fungorum Carpologia*, earned him the sobriquet "The Audubon of Fungi," in reference to John James Audubon, the American ornithologist and artist. Another Frenchman, naturalist and entomologist Jean-Henri Fabre (1823–1915), produced over 600 sensual yet scientifically precise watercolors of fungi that are housed in the archives of the Muséum Nationale d'Histoire Naturelle in Paris. German zoologist Ernst Haeckel (1834–1919) was both an extraordinary scientist and an exceptional artist. His scientific observations and artistic creations culminated in his book *Kunstformen Der Natur*, published in English as *The Wonders of Life*, which revealed the intricacies of small life forms. While fungi were not a main focus of his work, they are featured in a handful of lavishly detailed illustrations of stinkhorns, earthstars, lichens, and agarics. Straddling the worlds of both science and art, Haeckel's intricate illustrations are highly stylized yet technically precise. In *The Wonders of Life* he describes how interest in natural and artistic forms "depends for the most part, if not altogether, on their beauty—that is to say, on the feeling of pleasure we experience in looking at them."

Beauty in nature has long been represented by birds, orchids, and other attractive organisms, but handsome fungi are having their day. New Zealand became the first country to feature a mushroom on a banknote when the blue pinkgill, a species native to the country, featured on the 1992 edition of its $50 note. In 2016 it was reissued, and the mushroom was duly promoted from the edge of the note to a more prominent near-central position. Known as *werewere-kōkako* by the North

Island Māori, the mushroom's name arises from the similarly colored blue wattles (ornaments) on the throat of New Zealand's *kōkako* birds. The Māori explain the coloration as resulting from the bird rubbing its cheek against the fungus. The mushroom's representation on the banknote brings its existence into high circulation in the public realm.

The choice of the blue pinkgill for the banknote is perhaps not surprising. The fungus is a stunner both for its color and elegance. It accompanies other charismatic native fungi that traverse the globe on New Zealand postage stamps, and its widespread online presence attests to its magnetic allure. When mycologist Peter Buchanan from Manaaki Whenua—Landcare Research in Auckland ran a public poll to determine New Zealand's "national fungus," the blue pinkgill was the clear favorite. Yet Peter and the Landcare scientists chose a different species. They nominated the vegetable caterpillar for the dual reasons of it being the first scientifically described species in New Zealand and its significance to Māori people. However, as it looks like a scrawny brown twig, its charm is more contentious. What's more, like teatree fingers, the vegetable caterpillar is also a parasite. Its way of slowly killing its caterpillar host from the inside out could further reduce its appeal to some people. Scientific and cultural significances don't always align with the aesthetic sensibility of the public.

The image of the blue pinkgill on the New Zealand $50 banknote was taken by New Zealand wildlife photographer Rob Suisted. Rob stumbled across it in a forest of southern beech on New Zealand's South Island while out photographing for a *New Zealand Geographic* magazine story. He describes it as "an irresistible little blue magnet for photographers." Rob combines his scientific knowledge and photographic skill to create compelling images of wildlife and natural landscapes, yet he also recognizes the need to bring less visually appealing species and places into the spotlight. "We're typically drawn towards things that are charismatic or cute, but many worthy conservation subjects don't have these characteristics," he observes. "It falls on illustrators and photographers to find ways to make subjects relatable to wider audiences. That drives me to go the extra mile in finding fresh perspectives."

Photographing fungi, as with any subject, is an entirely subjective process. It's selective and impressionistic. Over my years of photographing fungi, I have largely shifted from creating scientific images of fungi that illustrate diagnostic features to more aesthetic portrayals. In moving from these more technical images to those that reveal less and leave more to the imagination, I hope to inspire rather than simply inform the viewer.

Finding a specimen that could be considered representative of a species lost its allure for me. Being "representative" means a mushroom shares a majority of features with other mushrooms of the same species. But given the extent of variation of a mushroom at different developmental stages and with exposure to varying environmental conditions, I realized that a single specimen could never be truly representative. It would be similarly difficult to represent the great diversity of *Homo sapiens* with just one image of a human being. I was more curious about the distinctiveness of each specimen than about what they shared. Being able to put a name on a mushroom from a clinical depiction of diagnostic features mattered less than simply being open to enchantment.

Today, some artists are deeply steeped in the mycology that inspires their work. Some even apply scientific methods to create mycological materials for their art, looking beyond the aesthetics and into fungi as artistic ingredient. In something of a new artistic paradigm, bio-artists explore fungi as subject matter and medium, metaphor and muse, working collaboratively across artistic disciplines and with scientists. Mycelium, for example, is not only weaving its way into fungal alternatives to leather, packaging, and building materials, but in the hands of artists it's turning up in sculpture, fashion, and furniture. It grows rapidly, and can be locally and easily produced in large amounts. It's also strong, it's cheap, and it has a low environmental footprint, offering an alternative art material.

Charismatic Aliens

The work of artists, as well as conservation tools such as flagship species, help put fungi in the public eye. However, being beautiful can also come at a price. What happens when an enthralling fungus becomes a potential pest? Aesthetic pleasures can become invasive. We might need to avert our gaze beyond their mesmerizing looks to see what they're actually doing.

The status of a fungus can switch from aesthetic virtue to environmental vice as understanding grows, ecological concepts change or a fungus relocates. Changing climates and environments, and help from humans, are enabling some fungi to expand their territory. Mycorrhizal fungi can colonize new places by hitchhiking with their tree partners or courtesy of humans who translocate them. Some fungi are deliberately introduced in forestry or horticulture to maximize the success of plant partners. It seems, however, that not all introduced fungi are staying put.

In Australasia, the fly agaric is an example of a mobile species with a changing status. With its striking red, polka-dotted cap and impressive size, it's probably the world's most familiar and most photographed fungus. For some people its mind-altering properties and reputation as an intoxicant heighten its appeal. Introduced to pine plantations in Australia from the 1920s, it's among the most conspicuous of the fungi that arrived in the country from elsewhere. It also forms relationships with introduced broad-leaved trees such as oak and birch.

Until recently, the fly agaric had stayed with its host trees in plantations, parks, and gardens. However, in southeast Australia and in New Zealand, it is moving from its introduced pine partners into adjoining native temperate rainforests, where it has hooked up with native myrtle beech trees. Once admired by the public for its aesthetic fairytale appeal, the fly agaric is now considered by mycologists to be an aggressive invader. Whether it displaces the myrtle beech's native mycorrhizal fungus partners is not yet known, but mycologists think it likely. Displacement of native fungi could not only diminish fungal diversity, but could render the myrtle beeches less able to tolerate stress and disease. With weaker trees and fewer native fungus partners, the overall resilience of these forests could decline.

The fly agaric has only been in Australia for about a century and has switched tree hosts in a relatively short time. With few mycologists employed to study the ecology and distribution of fungi, it's hard to know how this host-switch could play out, or indeed, what could happen if the fungus takes a fancy to *Eucalyptus*, Australia's dominant tree genus.

In New Zealand, the fly agaric has been classified as a regulated pest since 2001. Yet despite its pest status, as public interest in fungi grows, the popularity of the fly agaric soars ever higher. I've received queries from people as to how they can grow "that pretty red fairy mushroom" in their gardens. At least the fly agaric's conspicuousness and mycorrhizal reliance on tree partners means mycologists can, to some extent, keep an eye on its movements. On the other hand, saprotrophic fungi have the freedom to get around without being limited by a mycorrhizal partner. Saprotrophic fungi on the move present a bigger challenge to mycologists who are trying to keep watch on their whereabouts.

In the fall of 2012, I was hiking through Lilly Pilly Gully on Wilsons Promontory, at the southernmost tip of the Australian mainland. Its temperate rainforests and coastal heaths revealed a rich variety of fungi, but also an unexpected newcomer. On the side of the track I noticed an old log spotted with orange dots. On closer inspection, I saw that they were not the familiar blobs of golden jelly bells as I'd expected, but something I'd never encountered. Each tiny, fan-shaped cap was attached to the log by a stubby stipe. Their undersides were puckered with large tear-shaped pores, giving them an appearance a little like honeycomb. It was an exquisitely beautiful and delicate thing, resembling a similar-looking fungus with the whimsical name of the little ping-pong bat. This one, however, was a striking bright orange and, as I soon discovered, it is known as the orange ping-pong bat.

Occurring naturally in Madagascar and parts of southern

Asia, this fungus was first recorded in Australia in 2004 in a suburban Melbourne park. No one really knows how it found its way to Australia, but it may have been present in timber shipped from New Zealand. Today the orange ping-pong bat has established itself in hundreds of places across five Australian states, and continues to increase its distribution. Many of the locations in which it grows are considered ruderal—areas exposed to human disturbance, such as track edges and picnic grounds—pointing the finger at *Homo sapiens* as a highly effective vector of their spores. Three years after my first sighting, I returned to Wilsons Promontory and spotted it again on the other side of the peninsula, 20 miles from Lilly Pilly Gully. The fungus had somehow found its way over Mount Oberon, possibly transported by unsuspecting hikers.

In Australia and elsewhere in the world, this fungus is now regarded as a "fungal weed." Like the fly agaric, it has mycologists worried about its potential to compete with native fungi, especially given its ability to produce antifungal compounds to ward off fungus competitors. And being a saprotroph that grows on a wide range of plants, it is more difficult to track than the mycorrhizal fly agaric. In New Zealand alone it has been found on the wood of more than 50 different host species, both native and exotic.

A saprotroph's potential to spread is greater than that of a mycorrhizal fungus that is confined by the distribution of its host trees. Although fungi like the fly agaric have switched hosts, allowing them to claim more territory, the rate of dispersal is generally slower than for saprotrophic fungi. The orange ping-pong bat's sporing bodies may be tiny, but its distinctive and stunning appearance further complicate the scenario. For some people, this fungus proves irresistibly attractive, with the risk that they might collect it, relocate it, and unwittingly contribute to its spread.

The orange ping-pong bat highlights the challenge of negotiating aesthetics and ecology in conservation. Given it's both a charismatic species and a successful colonizer, there's a great need to understand its invasive capacity and its potential threat to ecosystems. Aesthetically beautiful flagship species may catalyze conservation, but "charismatic invasives" can hinder management efforts if people object to the control of species they perceive as beautiful.

Non-Conforming Stiltballs

Attractiveness can be about visual appeal or aesthetics, as well as other qualities such as rarity. But, like attractiveness, the notion of rarity can also be slippery. A species can be more attractive to people if it is rare, but determining rarity is easier said than done. It was in the desert country of western Victoria that a confounding fungus brought these conservation challenges to the fore.

Conservationist Sue Smith invited me to see an unusual desert fungus, the sandy stiltball, that I'd not yet met. It was raining heavily as she swung open the gate of Snape Reserve, adjoining the Little Desert National Park, and waved my car through. The windscreen wipers flapped madly as Sue leapt in, water pouring from the brim of her hat. As we sped along the track, a flock of bedraggled emus appeared alongside us, easily keeping pace until suddenly veering off into the scrub.

Snape Reserve was acquired by Trust for Nature in 2002. Sue having retired from teaching science and her husband

Lindsay from farming, they've been the backbone of the 2,090-acre private conservation reserve since its acquisition. The word "desert" often conjures images of endless red sand, windswept dunes, and scattered spinifex. However, the Little Desert, like most of the world's deserts, is vegetated. "Desert" simply refers to its low rainfall and the low fertility of its sandy soils. Despite its 160-year history of grazing and cropping before becoming a reserve, Snape harbors a dozen different vegetation communities, including the endangered buloke woodlands. Deserts are not where most people go to find fungi, and the reserve has never been systematically surveyed, but Sue keeps track of the comings and goings of everything in there, and few species escape her attention. She and Lindsay devote every spare moment to documenting species, managing weeds and feral animals, revegetating, and restoring.

"Head over there," said Sue pointing off the track. We cut through the sand and I pulled up near a desert stringybark just as the rain eased and weak rays of sun pierced the clouds. Sue opened the car door and jumped out before we'd reached a standstill. "There they are!" I followed, treading carefully and taking a moment for my eyes to adjust and recognize an unfamiliar species. Despite its size and striking appearance, the sandy stiltball blended remarkably well into the desert woodland environment. We knelt in the sand to have a closer look. "This one is so tough and woody, and it's persisted since last autumn," Sue tells me. "The others have just come up. It's the strangest thing!" The sandy stiltball is certainly an unusual fungus, impressive for its stature, growing up to a foot and a half tall on a shaggy, woody stipe. On top is a domed sac with a papery outer casing, inside of which is a powdery mass of rusty-brown spores. At maturity, the sac ruptures to release its spores from the height advantage provided by its unusually tall stipe.

As with desert fauna and flora, desert fungi are superbly adapted to cope with hostile conditions. The sandy stiltball's tough woody stipe allows it to withstand the extremes of temperature and abrasive winds of a desert life. It is found not only in Australia but in 64 countries scattered across every continent except Antarctica. It mostly grows in dry and sandy habitats, from steppes and deserts to savannas, coastal dunes, and woodlands, but is not restricted to these environments. In Bulgaria it grows on alluvial riverbanks, while in Hawaii it is found at altitudes of over 6,500 feet. In Macedonia it has been recorded on an offshore island in soils rich with the guano of cormorants. In Brazil it grows in subtropical rainforests. In 2017 it was found on Mount Victoria in New Zealand. Most likely it arrived there from Australia via the westerly winds that have also bestowed other fungal gifts upon New Zealand, including the plant pathogen myrtle rust. The sandy stiltball also pops up in habitats created by humans, from cemeteries to military training fields, road verges, and hedgerows. Its appearance in these ruderal environments suggests it can tolerate human disturbance.

Although it has a widespread global distribution, the sandy stiltball is recorded infrequently and sporadically, and on a global scale is considered rare. In Macedonia, Serbia, and Poland, it has disappeared altogether. It appears on the Red Lists of many countries, and is among the species being considered by the Global Fungal Red List Initiative. Assessing extinction risk is a complex and lengthy process, and whether a species is truly rare or wrongly or under-observed is hard to determine. Preliminary assessments suggest it will probably be listed as of "low concern" due to its wide distribution and appearance in disturbed habitats. As new information comes to light, species are reassessed. The changing whereabouts of fungi and their survival strategies throw up interesting questions about the nature of rarity, especially when a species that is deemed rare begins to flourish in human-disturbed habitats. When the sandy stiltball popped up in a pot of geraniums in Italy, it had researchers scratching their heads about its real status in that country.

The sandy stiltball relies on wind to disperse its spores. However, it seems to be benefiting from a long-distance vector in humans who supply potting mix and peat to the nursery industry in Italy. It has been surprising gardeners and horticulturalists by turning up in garden beds and pots of ornamental plants. These fungal aliens are nothing new. The name of another alien, the plantpot dapperling, suggests its preference for human-made habitats. Some researchers question, however, whether the sandy stiltball is as "rare" as its Red Listing in Italy might suggest.

This fungal alien among the geraniums highlights the challenges of assessing the rarity of species, especially as humans modify environments and the effects of climate change push them to new terrains. If a species is rare in its natural habitat but flourishes in human-modified environments, should it still be considered rare? And is its risk of extinction reduced, increased or unchanged when it moves in among humans? If you're a rare fungus, plant pots aren't exactly steady or reliable environments to colonize. Neither are road verges. In the United Kingdom the sandy stiltball is protected, but when an overzealous bulldozer driver decimated a known sandy stiltball verge habitat, the lack of consequences for his actions—unintentional as they were—exposed the weakness of protection laws.

Like the fly agaric, whose status is shifting from revered fairytale mushroom to invasive weed, the sandy stiltball in some locations could move from rare species to "potting mix contaminant." A species can be rare in its natural habitat while invasive in another. Fungi are mobile; they appear in many guises and are perceived in different ways, depending on who is doing the asking. Approaches to assessing rarity are constantly being reviewed and improved as more is understood about how and where fungi live and disperse. While single-species protection is necessary, the vagaries of the sandy stiltball highlight the challenges of conserving fungi, and reinforce the importance of protecting habitat at an ecosystem scale.

Setting aside areas for protection, such as national parks and other conservation reserves, has been a backbone of biodiversity conservation globally. While national parks play a vital role in biodiversity conservation, new research in Australia reveals that even large ones are failing to halt the decline of biodiversity and extinction of individual species. Although species might be protected within national parks, their home ranges often extend beyond the bounds of a protected area. National parks are necessary but are unlikely to incorporate the long-term and large-scale dynamics of ecosystems. They are not enough to slow or halt the decline of biodiversity, including fungi.

Today some of the best conservation, as demonstrated by Sue, Lindsay, and the Snape Reserve committee, is taking place on private land. Many landholders I've met who live on or regularly visit private reserves are acutely aware of the species with which they share it. They go to great efforts to protect them, often working in partnership with local communities and conservation groups, Traditional Custodians, and researchers. Many think beyond their border fences and take a whole-landscape approach to conservation. This is especially important for fungi that are less conspicuous and less well known than other organisms. Fungi are rarely surveyed or monitored—and without species inventories or "baselines," the extent to which environmental changes affect them are mostly unknown.

Taking Time for Noticing

What small, private conservation reserves might lack in size, they gain in the detailed attention to species from their trustees and volunteers. Two hundred and fifty miles south-west of Snape Reserve, near Apollo Bay, I stood on a bluff and gazed out over Bass Strait. Beyond the reef, Australasian gannets soared and wheeled before plunging into the ocean. Squalls of rain moved along the horizon in the fading light. In the paddock below, I saw the headlights of a pickup truck as it moved sporadically across the terrain. It slowly traveled east then stopped, reversed, and did a figure eight before it drove south, then stopped and reversed again. It seemed an unusual way to drive. I reached for my binoculars and realized that Jenny Robinson and Willie Bedford were out collecting mushrooms for their dinner.

Keen to see what they'd found, I grabbed my coat and tore down the hill and into the paddock, waving madly. Jenny saw me and sped over. "Jump in!" she yelled over the wind, and I landed in the back among the baskets of mushrooms. Lifting a large field mushroom out of a basket, I felt its solid, satisfying weightiness in my hand and admired its underbelly of perfect pink lamellae. Content with the bounty, we bumped along, checking the progress of newly planted seedlings and surveying the boundary fence for breakages. Field mushrooms pop up reliably in the paddock most falls, and while they make a tasty meal, the fungi, flora, and fauna are all protected in the rest of the 104-acre property that forms a private conservation reserve known as Seatrees.

For more than three decades, Seatrees' custodians— Jenny, Willie, and Pat Barnes—have observed the comings and goings of birds, mammals, mushrooms, and more; and have noticed, recorded, and documented everything. I'd been visiting the property for about a decade, and in 2017 was lucky to be appointed ecological adviser. Seatrees harbors diverse communities of vegetation and threatened species like the rufous bristlebird. With its varied habitats and abundant organic matter, it is also a haven for fungi. The trio closely tracks the habits and behaviors of the reserve's 110 bird species, the appearance of a great range of fungi each fall, and the migration of southern right whales along the coast. They amble slowly, notice details, and reflect on changes. When they discover a species not previously observed, they watch and research it to understand how it fits into the larger ecology, or if it has particular needs.

These sorts of daily, long-term, and meticulous observations are rare and seldom happen at the scale of national parks. Park managers and rangers have the huge task of juggling the demands of park visitors, managing facilities, and overseeing the protection of natural and cultural "resources," usually with pitiful budgets. They do what they can, but sufficient time or money are rarely allocated for the detailed surveying and monitoring of the species and ecosystems the parks are designed to protect.

Protecting large swathes of land that incorporate whole ecosystems is a priority for biodiversity conservation. Larger reserves usually offer more diversity of habitats and a greater area for larger populations. They allow for the movements and migrations of species, and hence a larger gene pool. Relative to many conservation reserves and national parks, Seatrees, like Snape Reserve, is small, but it should not be underestimated. Small reserves provide valuable corridors and stepping stones to other protected areas. They are easier to restore and conserve than larger reserves, as species can be more easily observed, monitored, and understood. Their scale makes them more manageable.

For fungi, habitat loss is the single most significant threat, as it is for flora and fauna. Clearing of coastal land for urban developments has destroyed and fragmented areas surrounding Seatrees. Ever-growing concerns about wildfire see vast amounts of organic matter burnt and removed every year in the surrounding Great Otway National Park, reducing fungal diversity and the health of the forests. Seatrees is covenanted for protection in perpetuity, and provides the kind of long-term sanctuary that is not guaranteed by national parks.

Disturbance of soils and vegetation disrupts fungi. Minimizing disturbance underpins Jenny's philosophy for managing Seatrees. She speaks of their unobtrusive and careful approach to learning about the reserve. "The thought of brutishly bulldozing any bush on Seatrees was never an option, so clearing the tracks by hand took the best part of a year," she explains. "This slow process provided time to notice all sorts of features, including significant trees, groundwater springs, and the nests of little birds." The native habitat of Seatrees is scrupulously maintained, and damaged areas are continuously restored, revegetated, and enhanced.

The morning after Jenny and Willie's mushroom collecting, we wander along a mossy track and there's a sudden thumping through the bush as a wallaby startles and takes cover. It darts beneath the tree ferns and disappears into the gully. A koala silently observes the commotion from above. The reserve also supports a registered wildlife shelter; Willie has nursed hundreds of injured animals back to health and released them. Along the track, a cluster of ocher-colored coral fungi poke through the leaf litter. Jenny drops to the ground, peers at it, and pulls out her camera to photograph it. From her ground-level vantage point, she spies others: the ruffled rosettes of the wine glass fungus hug the base of a eucalypt, and an emperor cortinar, in magnificent purple, is just starting to reveal itself. Jenny notices and records everything. Nothing escapes her gaze.

Conserving fungi is challenging. It may be founded in scientific understanding of species and ecosystems, but it is driven by humans working on the ground.

Red Lists and Bucket Lists

Many of us live in increasingly digitized worlds. Noticing the local can sometimes give way to the desire for constant global connection. The chance to witness something miraculous right before us—perhaps an emerging basket fungus or Jenny's little birds' nests—can be overwhelmed by the fear of missing out on something on the other side of the world. The global, it seems, has become more pressing and prestigious than the local. The digital revolution has enabled us to scale up the data we collect about nature and submit them to giant online repositories where they're plugged into ever more planetary algorithms. This living census of the world's species contributes to scientific understanding of biodiversity, but while we're frantically uploading our images from our phones, we can also pause to notice what's beneath our feet.

Photographing beautiful things and sharing those impressions is human nature, and has been critical to conservation, including the conservation of fungi. The aesthetics of nature was an important driver of early conservation movements, and it remains tightly woven into conservation today. Decisions to create the first national parks, especially in the United States, hinged on an area's perceived scenic or picturesque qualities. However, as understanding of ecology has grown, so has the tension between aesthetics and science in conservation. Just as traditions of preserving charismatic species can override ecological values, so can the push to conserve only areas of scenic beauty. Places that have higher ecological values but less aesthetic appeal can be overlooked. The survival of fungi that live in commonly undervalued habitats, such as grasslands or coastal scrub, hinges directly on those particular ecologies and conditions. Yet these habitats are often among the first to be cleared for farming or development, especially if coastal views are on offer.

The notion of the picturesque once described places considered scenic or view-worthy. Today, the picturesque has morphed into those places and species considered "Instagrammable." The geotagging of photos means that locations once less known or remote have become easier to identify and access. For some keen snappers, Instagram is a new guidebook for global travel. However, critics worry that people who post images with little concern for the subjects they photograph, or the impacts of their own presence, are threatening both species and environments.

It was just before 5 a.m. one late summer day in 2019 when ranger and photographer Erik (not his real name) pulled up in his four-wheel drive in the parking lot of Fjaðrárgljúfur Canyon in southern Iceland. I was about to head off in search of fungi when he leapt out of the vehicle and stormed toward me, hat pulled low, bristled jaw jutted firmly outward. Erik locked me in his myopic glare and barked a series of questions that came out as accusations about where I'd camped overnight. He was clearly not in a good mood. He took a bit of convincing, but finally he accepted that I'd made an early start and arrived from elsewhere, and had not camped illegally in the parking lot.

Erik and I then engaged in a long and disturbing conversation. He told me of his battles to manage what he called "the spike of Instagram-obsessed tourists," and how he was greatly concerned about the damage being done to the local ecologies and fragile soils. "See all this bare ground?" he said, sweeping his arm in 180-degree arc. "This area was covered in lichens and mosses. Now they're gone, in just a few years. Clambered on and crushed. They take decades to grow and will take decades to come back, if they do at all, because our soils are unstable and growing seasons in Iceland are short."

Iceland experienced an unprecedented boom in tourism that helped lift the country from economic recession following a devastating financial crisis in 2008. However, Erik, like other Icelanders I spoke with, lamented the lack of infrastructure necessary to protect vulnerable ecologies. As both a photographer and a ranger, he was acutely sensitive to the interplay of aesthetics and conservation: "Images of beautiful places are consumed like junk food through Instagram. There seems to be little conscience or conservation awareness." The images are "just trophies," he said, throwing his arms in the air. "We can't build boardwalks fast enough to cope with the stampede of Instagrammers." I thought about the tragic irony of how places and species once revered for their beauty are being destroyed by the same aesthetics that used to protect them.

And it's not just happening in Iceland. Tasmanian-born hiker Rima Truchanas told me how the iconic places her photographer father, Olegas Truchanas, once photographed to showcase their beauty and need for protection are likewise succumbing to the increased influx of visitors. Olegas's photos of south-west Tasmanian landscapes were pivotal to the rise of ecological consciousness and conservation in Australia that gained momentum in the 1960s and 1970s. Yet today, that same ecological consciousness is often over-ridden in the rush to capture and post images of Tasmania's places online, as a confirmation of having been there, and to be rewarded in the currencies of "likes" and "followers."

As photographers, Erik and I discussed how photographs can stimulate and inspire interest in nature, but also risk reducing it to something to be simply "looked at" rather than something to which we are inherently connected. It's a line that's hard to draw. Capturing an impression on the sensor of a camera inevitably strips the subject of all but the visual. We talked about the paradox of how the "trophy hunting" of animals with guns has largely been replaced today with cameras, yet "shooting" with cameras also has environmental consequences. "Cameras don't shoot bullets," said Erik "but the scale, the sheer number of people out here taking pictures, is the issue. I think hunters understand their responsibility and the implications of their actions better than many Instagrammers. But what can we do?"

The effects of careless photographers on some high-profile species have drawn attention. Puffin chicks in clifftop nests crushed under nonchalant selfie-takers' boots have raised the ire of Icelanders. But I wondered how less well known organisms like fungi might too be affected by the increased tourism, especially off-road driving, about which ranger Stefanía Ragnarsdóttir has expressed her concerns. Rare fungi grow in Iceland. Several are listed on the IUCN Red List of Threatened Species. Grassland waxcaps—such as the blushing waxcap, crimson waxcap, nitrous waxcap, dingy waxcap, and gray waxcap—are all rare and listed as vulnerable. They could well disappear under the wheels of too many off-road vehicles. Or even one vehicle. Icelandic mycologist Guðríður Gyða Eyjólfsdóttir told me that the violet coral, a rare and vulnerable species, has only once been recorded in Iceland, while the mealy pinkgill has only ever been found on a single hillside. Yet, like elsewhere in the world, there are few resources to survey and keep track of vulnerable fungi.

Erik took off his hat and smoothed back a mane of blond hair. He looked deeply tired. His concern about the environmental damage to his homeland was taking a profound personal toll. After we said goodbye and I headed down the trail, I thought about the different ways we learn about nature, especially how many indigenous cultures transfer knowledge. Their knowledge of ecosystems and species is rooted in cultural experience and strongly contextualized to place. The rituals and responsibilities involved in knowledge transfer have to be earned. Recipients are entrusted to protect traditions about knowledge, and transfer it in accordance with cultural customs and lore. Underpinning the transfer of knowledge is the shared premise of our inherent connectivity with nature. To disregard any part of nature is to disregard the whole, with direct effects through space and time. To stand on just one puffin chick, or one fungus, has implications. The gaping divide between this approach and the "trophy hunting" Erik described loomed large in my mind.

While the commodifying of nature is not new, as Erik warned, the scale is unprecedented. Instagram and other social media platforms allow for the instant global dissemination of images and location of species and places, but they're not the only source. Online repositories of biodiversity data similarly provide geographic information on the whereabouts of species, including endangered ones. Such platforms have inspired public interest and engagement in biodiversity conservation. Public involvement in the mapping of species contributes to our understanding of how rare species are, how they are distributed, and how their populations shift. It is an important part of conservation. The rapid spread of information via social media can also be a great boon for reaching wide audiences, especially when rallying support or when an urgent response is required.

However, like things of beauty, rarity is a drawcard. Although fungi are more likely to be sought for their culinary or medicinal value than their rarity, some mycologists who develop fungus distribution maps question whether it's wise to make the distribution of rare species publicly available. Sapphire McMullan-Fisher tells me how she deliberately keeps the locations of the teatree fingers fungus ambiguous when putting distribution data online. "People want to photograph everything, especially rare species," she says. "Teatree fingers grows in heath and thickets, and trampling can have an enormous impact in these ecosystems. There's also the risk of introducing species from elsewhere that can upset local ecologies." She describes how it's not just rare fungi that attract attention, but fungi with rare traits. She reels off a list of typical species on photographers' "bucket lists," many of which have unusual colors, including some that glow in the dark: "Blue and luminescent fungi are uncommon so therefore very popular, and local sightings often cause a photographic flurry. But without care for their habitats, the delight in our fungal treasures could threaten their survival."

I pondered Erik's and Sapphire's concerns. While a picture supposedly paints a thousand words, I wondered if we pause long enough to consider each one, as they endlessly scroll past our eyes. But despite the threat of practices like unscrupulous Instagramming, for many people, the natural world becomes richer and more vibrant with each species observed, appreciated, or photographed. Perceptions are challenged and imaginations stretched. I thought of Bea Harris and how the beauty of the verdigris agaric was in itself enough to make it meaningful to her. Yet understanding fungi can enhance our aesthetic appreciation, and endow them with another kind of meaning.

Public perception and aesthetic appreciation of fungi constantly throw up new challenges and possibilities. Charismatic fungal flagships have increased awareness of fungus conservation, but recognizing the diverse aesthetics of fungi is part of knowing and valuing them. Some ordinary-looking fungi do extraordinary things, concealing interesting backstories that only come to light in the context of their environments or in their hidden details beneath the microscope. Real progress in fungal conservation happens when it moves beyond the beautiful to consider fungi in all their bizarre manifestations.

9

WOMEN AS KEEPERS OF FUNGAL LORE

I reach into the undergrowth to pluck a perfect fat porcino only to feel a sharp pricking pain in my finger. Pulling back my hand with a yelp I realize it was a sneaky trick for a porcino to pop up among an armory of spiny sweet chestnuts. My Swiss hiking companion, Valérie Chételat, looks on, amused. She's the one who knows the way. I'm supposed to know the fungi, but no one told me Swiss ones bite. We're wandering the winding stone paths and mule trails of Monte Bigorio on the border between Switzerland and Italy. They're strewn with swirling leaves of beech and oak and—as I've just discovered—piercing chestnuts.

Twists of cloud snake through the valleys, offering momentary glimpses of the mountains. They're dressed in bands of fall color: golden larch and copper beech, yellow birch with white skeletal trunks, and the green tones of spruce. Shafts of light warm the forest floor, releasing the resinous scent of conifer needles. We climb old stone steps, past wayside shrines and upwards through abandoned settlements. Their inhabitants are long gone, leaving during the economic hardships of the late 1800s. Some moved down to the valleys. Others sought prosperity further afield in the gold rushes of Australia and America. Now only goats seek the shelter of the crumbling ruins. And no one, it seems, is out seeking mushrooms.

The absence of foragers is curious given the sumptuous bounty of fungi. We spy golden chanterelles hiding in the leaf litter. They're a favorite throughout Europe. In Sweden it feels like they hold more value than the krona; one should never get between a Swede and her chanterelles. That said, I'm not taking my chances with the Swiss either. A little further along, hedgehog mushrooms, with their stalactite-like undersides, hug the ground. The smooth black trumpets of another chanterelle, the horn of plenty, scallop the track edges. We climb onto a grassy plateau strewn with majestic parasol mushrooms the size of dinner plates. They sport snakeskin stipes and we run our fingers over their felty caps. The forest is flush with keenly sought edible species. Then, as if to keep us on our toes, a deathcap looms in the leaf litter beneath a European oak. True to its common name, this mushroom can kill the uninitiated forager. Yet it is elegant and regal, with its pale-green satin sheen, and not remotely ominous.

We stop to photograph the deathcap and pause by a recently fallen birch. Its surface is puckered with the rubbery blobs of black witches' butter. Valérie squats down and peers at it closely. "It looks so benign," she says. "Yet in European myths, if you found it on your gatepost, it was a sign you'd been placed under a witch's spell!" Others assert it has the power to neutralize witchcraft, apparently when the fungus is thrown into a blazing fire. We agree that neither the spell nor the remedy sound too convincing. The softness of the bark on which this one grows suggests the fungus is more focused on finding a feed than invoking evil.

European folklore is rich in associations between witches and fungi. Witches abound in the common names of some fungi across languages. Germans refer to various *Hexenröhrlinge* (witches' boletes). Italians know the red cage fungus as witch's heart (*cuore di strega*), as do the French (*coeur de sorcière*). One of the old Icelandic names for the spore-releasing stage of the brown puffball is *kerlingareldur*, meaning "hag's fire." The image of a miniature cackling witch stirring the pot of the spent puffball, its black "smoke" of spores rising ominously, comes to mind. A more fitting vessel is perhaps the witches' cauldron, as it's known in English, with its dark-brown, tall-sided cauldron-like form.

Unsurprisingly, things foul or unkempt were also pinned on witches. Some English speakers use the name "witch's eggs" in reference to the egg-shaped immature stage of stinkhorns. Then there's a lichen with its wildly straying strands known as witches' hair. Even the innocuous-sounding fairy rings—mushrooms that grow in circles—have darker associations. Originally spelled *faeries* in English, these were capricious creatures of ill intent. The French, Germans, Italians, Swedes, and Icelanders refer to a circle of mushrooms as a fairy ring but also as a witch's ring (*rond de sorcières*, *Hexenring*, *cerchio delle streghe*, *häxring*, and *nornabaugur*).

Associations with witches go beyond names to include superstitious practices. In some Balkan countries, dried mushrooms were assembled as amulets to ward off witches. As the sun slips lower, I glance across the Val Colla at the strangely irregular dolomite mountain peaks of the Denti della Vecchia. They poke out at odd angles with gappy intervals. The name translates to "teeth of the old woman." I try to imagine them in the mouth of a giant grandmother as they are swallowed up in cloud, but can't be completely sure they aren't in fact *denti del vecchio*. Perhaps it was simply a spelling mistake that put them in the mouth of an old woman and not an old man.

From Fungal Wisdom to Fungal Curse

Women are thought to have been the crucial keepers of fungal lore across centuries and cultures. They observed where and when fungi grew, and determined which could cure and which could kill. Some philandering husbands are thought to have met their deaths with a meal of deadly mushrooms courtesy of their wives, although this may well be speculation. We can only assume women's knowledge arose from extensive long-term observation and experimentation, passed down through generations. However, as the lives of women were seldom documented, it's not known for sure how they gained their knowledge of fungi.

Women have traditionally been carers and healers. They included midwives and those who prepared sophisticated herbal medicines, and those who nursed the ill in homes and hospitals, and examined corpses for causes of death. Some, like Yolŋu woman Alison Djawurrku Wunungmurra from East Arnhem Land in northern Australia, continue to use fungi medicinally today. Over time, women developed not only a profound understanding of fungi, but the power that comes with knowledge.

The European Renaissance of the fourteenth to seventeenth centuries brought rapid and radical changes in the way knowledge of medicine was acquired and practiced. It was marked by a spirit of inquiry, the challenging of ideas, and the empiricism of the scientific method. Experimentation and dissection revealed new insights into the workings of the human body. Technologies such as the microscope enabled new discoveries, and the invention of the printing press saw their publication and spread. Under the new "professional" system, which required formal training (open only to men), licensing, and occupational titles, the traditional medical practices of many women healers and knowledge-keepers were denigrated and outlawed. Those who resisted and continued to practice risked being accused as witches. Many were submitted to unjust trials, tortured into confessions, and publicly drowned, decapitated, or burned alive. Men were persecuted in witch-hunts too, but fewer were put to death. Generations of traditional and mostly oral knowledge of fungi vaporized on the pyre along with those who had only sought to help others. Today, historians and anthropologists sift through scant records and clues as they attempt to retrieve this knowledge.

Theories about what led to the widespread hysteria that gripped early modern Europe abound. Witch-hunting was most extreme during the religious wars between 1580 and 1630. Witches may have served as convenient scapegoats for disasters, epidemics, and other inexplicable misfortunes. The changed climate brought on by the Little Ice Age (c. 1550–1860) caused radical disruption to social, political, and economic structures, and may also have contributed to the frenzy. While the cooling happened in phases, the most sudden and severe coincided with the height of witch-hunting. Not only was it colder, but temperature and rainfall rapidly and drastically fluctuated. The colder climate led to shorter growing seasons. Crops failed. Soils washed away. Disruption to the grain harvest caused widespread famine, disease, and death.

Tightly linked to the damp and frigid climes, a new protagonist entered the scene—a fungus known as ergot. The fungus grows on grains, especially perennial rye. A cold winter followed by a moist and cool growing season favors its flourishing. Some historians believe a wave of ergotism (ergot poisoning) could have catalyzed the witch hysteria. Ergot replaces shoots on grain with purple-black growths called sclerotia. When the contaminated grain was harvested and made into bread, its poisonous alkaloids survived the baking process, causing ergotism in those who consumed it. Symptoms of ergotism can be physically dramatic, including violent muscle spasms, and twitching of the limbs, tongue, and facial muscles. Some people are racked by strange sensations described as feeling like ants crawling under the skin. Others suffer gangrene of the extremities. Many are tormented by delusions and hallucinations. Such radical and frightening symptoms, along with misunderstanding of the disease, led doctors to diagnose patients as having been bewitched. Without the medical and mycological understanding of today, the belief that sufferers were witches may have seemed as good an explanation as any.

In the tranquil setting of the autumnal forests of the Swiss-Italian border, it seems inconceivable that between the fifteenth and eighteenth centuries, possibly as many as 100,000 Europeans were executed, and that a fungus might have had a hand in it. More than two centuries after the state-sanctioned murder of Anna Göldi, an innocent Swiss woman tarred with the title of "the last witch of Europe", Göldi was finally acquitted. She was one of some 6,000 Swiss who lost their lives to the witch-hunts.

Edging into Enlightenment Science

Religion and superstition gradually gave way to rational thought and science in the eighteenth century. Known as the Age of Enlightenment, this cultural and intellectual movement in Europe and America sought to reform and advance society through education. Enlightenment thinkers challenged the old order, casting off religious dogma and ushering in ideas of liberty and progress. Yet superstitions took a long time to shake. While attitudes and values were shifting in the private circles of Europe's elite, fear of witches lingered among the masses despite "witchcraft" having been decriminalized.

Although the Age of Enlightenment transformed thinking and social relations, things were slow to change for women. They had no political, legal, or economic rights. Most were denied education beyond what equipped them for household duties. To proclaim women as more "intuitive" and therefore incapable of intellectual or rational thought was a convenient way to keep them out of public life and confined to the drudgery of the domestic sphere. Women's curiosity or interest in science (or pretty much anything outside the home) was seen as a distraction from their household responsibilities. Some women resisted their oppression by exploring scientific and intellectual ideas through the rise of "salon culture," especially in France, Germanic Europe, and the United Kingdom. Salons were social gatherings of intellectuals who met for discussion in private homes, and hence women were able to attend. The rise of literature, especially fiction, during this era also gave women access to intellectual thought. By participating in literary discourses and artistic pursuits that converged with the scientific arena, women edged their way into the male world of science.

Women's lack of access to public life, however, remained a huge impediment. Universities and scientific societies were largely closed to women, depriving them of knowledge and interaction. A woman's access to scientific circles usually hinged on her association with a male relative, mentor, or spouse. As with most knowledge production in former centuries, women were not regarded as producers of knowledge in scientific fields, including biology and mycology. Yet many women enabled knowledge production through their contribution. Women were often collectors of fungus specimens for male scientists and researchers, and their work often went beyond collecting to include describing, illustrating, and preparing specimens. Others worked as translators of scientific texts.

Angelina Fanny Hesse (1850–1934), for example, worked as an unpaid illustrator and assistant to her husband, Walther Hesse, who was laboratory technician to microbiologist Robert Koch in Dresden, Germany. In 1881 she made a pivotal contribution to microbiology in proposing agar as a culture medium for growing and isolating bacteria, for which Koch took credit. Many women were unpaid and seldom acknowledged for their contributions to science. They served merely as ancillaries in supporting the knowledge creation of men, usually from the confines of the domestic space. There were, however, exceptions. Some women managed to not only penetrate the male domains of science but were recognized for their prominent and prolific contributions, including in the field of mycology.

Mycologist and plant pathologist Elsie Maud Wakefield (1886–1972) headed mycology at the Royal Botanic Gardens in Kew for nearly 40 years. She illustrated, described, and named many new species and, in 1929, was elected president of the British Mycological Society. Another pioneering woman, plant pathologist Johanna Westerdijk (1883–1961) was The Netherlands' first female professor. She researched and published extensively on plant pathology and mycology, and devoted herself to improving women's opportunities in science. Marie Schwarz (1898–1969), a student of Westerdijk's, isolated a fungus from diseased elm trees; another of Westerdijk's students, Christine Buisman (1900–36), determined that it caused Dutch elm disease. In Australia, Flora Campbell (1845–1923) was a pioneering mycologist who actively collected, illustrated, and annotated type specimens of at least 83 new taxa, making significant contributions over four decades. Yet prevailing prejudices and barriers saw her limited to honorary positions. For most women who had an interest in science, it was extraordinarily difficult to pursue scientific careers.

The Blob Taxonomist

Many women continue to work in honorary positions today, "honored" but not paid. Yet for those who have the time to devote to research, and who are able to work without payment, there are advantages beyond the fiscal. I asked mycologist Pam Catcheside about her honorary research position at the State Herbarium of South Australia, in Adelaide. Pam described how much she enjoyed the freedom and lack of pressure in her work. She pursues her mycological passion with the benefit of access to the herbarium collections and resources, but without the obligations and expectations of a paid position. Pam works longer hours than many people who are paid for their work, then comes home to write up her findings, respond to mycological queries, and work on various book manuscripts.

The understanding of where fungi grow and how they respond to changes in their environments often comes from the observations of honorary researchers and so-called "amateurs." With many fungi yet to be described and named by science, it is not unusual for undocumented species to be discovered. For some people, the idea of naming a species is romantic and exciting. However, once a fungus is collected, there's an extensive and systematic process to get a specimen named and accessioned in (acquired by) a fungarium. Naming a species involves the exacting and painstaking protocols of the scientific process. Few people have the taxonomic expertise. Without taxonomists, many aspects of mycology are not possible.

Pam is a deeply passionate and committed fungus taxonomist. The quintessential quiet achiever, it took a bit of needling from me to learn that she has discovered more than 20 new species, submitted 5,000 specimens to the herbarium, founded fungal studies groups and won a string of awards for her work. Pam is interested in all fungi but specializes in cup and disc fungi. These fungi appear as small cups or discs, lumps or blobs. Herbarium curator Helen Vonow refers to them as "Pam's little black blobs," although Pam informs me that she does "orange blobs as well." Black, orange or otherwise, Pam's blobs are mostly impossible to identify without microscopic examination of more nuanced and often stunningly beautiful characteristics such as spore ornamentation. Some spores are spiny, some are ridged, others are reticulate, and the spores of one species— *Sphaerosoma trispora*—have a beautiful double network of raised polygons.

Naming fungi, says Pam, "gives them respect and identity." She recognizes how vital they are to functioning ecologies, and the different roles they play. "They're all out there doing something excellent in some way, something really important," she says. "I like to see them and tell them they're doing a really good job!" Pam not only offers "respect and identity" to fungi but, through her example, to all the women who contribute to mycology. Her 22 years of commitment have all been in a voluntary capacity. Now in her ninth decade, Pam appears to be working harder than ever and, like the fungi, is doing more than "a really good job," enlarging taxonomic knowledge of fungi and inspiring the interest of younger generations.

The Wrong Chromosome

As an aspiring young naturalist, I'd known about the prestigious Linnean Society of London in a faraway land called England since early childhood. Bearing the name of Swedish biologist Carl Linnaeus and founded in 1788, it is the oldest active biological society in the world. I'd heard David Attenborough speak about it and knew it was where Charles Darwin's and Alfred Wallace's revolutionary theories of evolution and natural selection were first aired. In my imagination it was a melting pot that nurtured the most significant discoveries and ideas about the natural world. Several decades later, I was thrilled to be invited to speak on fungal conservation at a mycological meeting at the Linnean Society. Sitting in the audience beneath the stern portraits of bygone male scientists, I found myself wondering why, for the first century of its existence, the society had deprived itself of the contributions of women.

One of its early female candidates for membership was Beatrix Potter (1866–1943). Before creating the world of Peter Rabbit and company, Potter was an amateur mycologist, researching, experimenting, and creating hundreds of paintings of fungi and bryophytes. As women were prohibited from the society, she was unable to attend and hear the delivery of her paper, "On the Germination of the Spores of the Agaricaceae." It was read to the audience in 1897 by the then president of the British Mycological Society, George Massee, as was the accepted practice of the society at the time: that an author's paper be read by a proxy. Much has been written about Potter's life and the significance of her mycological contributions. However, critics have identified mistranslations of her scientific theories by her biographers. Some mistakenly credit her with the insight of recognizing lichens as symbioses between fungi and algae. Rather, her diaries suggest she held the more common and erroneous view shared by mycologists at the time that lichens were "standalone" organisms. The importance of Potter's contribution to mycology may have been overstated, but had she been given access to the resources and opportunities of her male contemporaries, the situation might have been otherwise. Potter went on to not only become a visionary conservationist, bequeathing 4,000 acres of her land to the National Trust, but an enormously successful and much-loved author, apparently reaping an income of which her mycological contemporaries might only have dreamed.

Women's membership of scientific societies was important not only in the fight for equal status, but for access to libraries, specimen collections, and other resources necessary to their research. It was the persistence and years of petitioning and campaigning by botanist and women's rights activist Marian Farquharson (1846–1912) that finally saw women accepted as members of the Linnean Society. When in 1904 the Linnean Society reluctantly conceded equality, Farquharson was the only one of 16 women candidates to be denied membership. The reasons for her rejection are unclear. Some historians suggest her tenaciousness and strength of will—qualities that made her such a successful campaigner—could have threatened those society voters who denied her membership. In 1908, Farquharson's application was finally reviewed, but she died before her membership could be formalized. It was another five years before women were granted permission to attend meetings or partake in discussions, and a further 60 years until a woman, Irene Manton (1904–88), was elected as society president in 1973.

Forty years later, I gazed around the meeting room to see that women made up half the audience. Professor of fungal ecology Lynne Boddy then took the podium. Lynne is a captivating and passionate mycologist who has pioneered work on wood-decay fungi and their interactions with invertebrates and bacteria. Her ground-breaking studies reveal how mycelial networks signal, communicate, and endlessly remodel themselves in response to changes in their environments. Lynne is a scientist who straddles the cloistered realms of research science and the public world of conveying scientific discoveries, and she recognizes how networked systems of fungi can also teach us about human societies. As Lynne captivated the audience with her animated descriptions of fungal warfare playing out within the clandestine depths of a rotting log, I felt like a child being told a terrifying yet thrilling bedtime story.

That evening, I visited the website of Lynne's fungal ecology laboratory at Cardiff University in Wales. The faces of nine women (one with electric-blue hair) and one friendly-looking chap beaming out of my monitor were a reassuring reminder that things are changing for women in mycology. That is, for some. In countries where women and men are given equal access to education, women are excelling in the biological and environmental sciences. Elsewhere, women have been flung back into the dark ages. Since Afghanistan's government collapsed in 2021, Afghan women have been deprived of education and personal freedom, and their futures remain horribly uncertain. Women comprise more than two-thirds of the world's 796 million people who cannot read and write, according to the United Nations. There's still a long way to go.

Lynne has worked in mycological research for almost five decades. I asked her how the experience had been for her as a woman in the field of science. Lynne is warm and unhurried, and seems to exist in a perpetual state of amazement about the curious lives of fungi. She paused, with a hint of puzzlement, then said, "Well, as a child, I could kick a football better than most of the boys." It said a lot. When she'd been told she'd have to be better than men to succeed in science, she'd just shrugged. I got the impression Lynne has been so focused on her fungi that she's had little time to ponder the challenges for women in science. She has inspired and motivated generations of young scientists, both women and men, and has forged ahead undeterred. "But yes, in retrospect," she added, "I think it's been tougher for women than men to excel in the sciences."

Kicking Goals for Fungi

Revolutionary ideas almost always attract resistance, and speaking truth to power can come at great cost. A mycologist working on mycelium and its wider implications may break new scientific ground while not making too many waves. A mycologist working in a forestry industry driven by ruthless economic imperatives is another story. But occasionally there are those dissenters who are heard, who are accepted, and who revolutionize thinking in their fields. One mycologist who had a tough time having her revolutionary discoveries acknowledged is Canadian forest ecologist Suzanne Simard.

Over the last three decades, Simard has challenged conventional wisdom about how trees, fungi, and forests function. Through years of extensive field trials she demonstrated how interactions between trees are not only competitive but also cooperative. Simard's observations of carbon transfer and resource distribution between trees suggest that relationships are more sophisticated than our scientific premises and forestry frameworks allow. Her discoveries contested one of the most fundamental principles upon which Western understanding of nature is built— that of the notion of competition in nature. Simard was only in her thirties when her findings were published in the prestigious *Nature* journal. Her claim that trees collaborate met with ridicule and criticism. The notion of competition is a powerful one. It is a foundational concept in ecology, and has infiltrated other fields including forestry, horticulture, and agriculture. The idea that fungus–plant relationships beneath the soil are more complex than previously known expanded understanding of how forests function, and raised important questions about how forestry is practiced.

Simard not only challenged a firmly cemented scientific idea, but she did so in fields (forestry and science) that are traditionally the domain of men. Adding to the ire of some critics was that she dared break another sacred rule of science: she anthropomorphized (ascribed human attributes to other species). Simard's notion of "mother trees"—large old trees that act as hubs for exchanges between plants and fungi—attracted heavy criticism. Some scientists warned that while altruistic notions capture the popular imagination, personifying organisms breaks scientific protocol and raises important questions about how scientific ideas are relayed to the public. Others criticized her for inflating the importance of cooperation while overlooking competition and promoting what they considered a misleading view of how forest ecosystems function.

However, Simard doesn't use metaphors lightly. As a scientist of great rigor and integrity, she is only too aware of the dangers of anthropomorphizing. Simard describes how such concepts are well established in the oral histories and writings of the First Nations people of North America's west coast. She sees it as a shortcoming of science to overlook what has been understood for thousands of years by indigenous peoples. "Western science is really just the little sister to aboriginal science, which has been going on for millennia," she observes. Simard argues that there is more happening in forests than can be detected through the limited tools and rules of Western science: "We've been so narrowly focused on reductionist science, pulling things apart and then trying to understand them, that we lost our way to actually see these as whole systems." Adopting less technical and more familiar language, says Simard, helps people relate to the interrelations of species in forests, and encourages care and stewardship.

The backlash Simard endured for challenging ideas about how forests function seems harsh. In her book *Finding the Mother Tree*, she details her relentless battles to convince her critics of her findings, and the efforts of some opponents to foil her research. Simard's work also prompted important ethical questions. She asks what it means to live, that is, to thrive, rather than to merely exist or survive. Simard exposes how prevailing forestry models push forests to their absolute limits, surviving at their leanest and least resilient. "When I look at forestry practices or agriculture practices or fishing, it's like we manage them just to survive," Simard says. "We don't manage them to flourish. We push them to the brink of collapse [and] take as much as we possibly can."

Fungi are the quiet achievers of the connections established by Simard and other mycologists before her. However, the conservation of fungi requires more than mycological knowledge. It requires being able to convey the science to wider audiences. Women are excelling not only in the science of fungi but also in their promotion and conservation. Amy Mullins is a radio presenter on Melbourne's independent radio station 3RRR. Her program, *Uncommon Sense*, covers a range of themes, from politics and international affairs to the natural world, but Amy has always had a soft spot for fungi. I asked her how she became interested in fungi, and in her enthusiastic response she barely drew breath: "Bringing an end to native forest logging has always been a huge concern of mine. To understand forests better, I became more and more interested in soils and soil health." Amy then listed the ecologists and mycologists, farmers and authors, and others she has interviewed who inspired her and influenced her thinking. "I started to appreciate the mysterious connections and communication of the 'Wood Wide Web' and how it all fits together," she explained.

Talking to Amy, I'm reminded of the power of radio to reach broad audiences. Those of us working in conservation often talk to like-minded people but can struggle to attract the interest of politicians, policy makers, and funding bodies which are pivotal to its success. Amy goes to great lengths to understand the science and convey it to her audiences in exciting and accessible ways. "On my show I like to talk about things that people may not fully appreciate but are hungry to know about and understand. I've built up a listenership around these subjects, and I hope the feeling of wonder that I experience is also felt by my listeners," she said. "My listeners are so enthusiastic about fungi. Often, the biggest response I receive from my audience on any topic is when I do interviews about fungi. Fungi seem to have attracted a whole new level of superfans!"

Making Mycological History on Mount Canobolas

Australian women have actively participated in conservation since before the Australian colonies were combined as states or territories in a single federated nation in 1901. They've contributed as scientists and explorers, artists, organizers and activists, polymaths and philanthropists. Over the decades I've worked on the ground in Australia with inspiring women who take leadership roles in conservation and especially in recognizing the importance of fungi in functioning ecosystems. Fungi have a dedicated and tireless ally in fungus enthusiast and Liz Davis, a facilitator with the farmers' conservation group Landcare.

Liz captures the groundswell of interest in fungi by organizing fungus forays and workshops, drawing together Traditional Custodians, Landcare volunteers, field naturalists, government employees, and the wider public. She collaborates with mycologists and conservationists to bring understanding of fungi to the people. Over the years, Liz and I have discussed how one of the first steps to conserving fungi is to rally awareness and interest. The next is to get a feel for the local species, and a survey is a good starting point. Liz didn't muck around. With a clap of her hands, she shot into action. In no time she had coaxed fungus enthusiasts from near and far to give up their weekend to search for fungi.

So, on a chilly late-fall day in 2021, I was looking forward to working with Liz on the first public fungus survey of Mount Canobolas, near the town of Orange in the Central Tablelands of New South Wales. The temperature gauge in my car read 26 degrees Fahrenheit, dropping further as I ascended the mountain. It was not yet winter and I was hoping the cold snap had not spelled the end of the mushroom season. My eyes were habitually drawn off the road and into the forest, scanning for fungi. A reassuring clump of spectacular rust-gills hugged the base of a manna gum tree, its smooth, strong trunk disappearing into the fog. Mustard-colored goblets of austral forkgills and great gangs of webcaps adorned the road verge. The forest slowly awoke with the eerie whooping calls of pied currawongs and the red flash of a scarlet robin. I had a feeling it was going to be a good day.

Mount Canobolas is a biological oasis, an island within a sea of agricultural and forestry monocultures. Set aside as a state conservation area, the mount harbors an important remnant of distinct montane and subalpine vegetation. It's unusual in that it supports a diversity of plants species, including those that grow in arid, alpine, and temperate rainforest environments. Nestled among the rocky outcrops is a community of endangered rock-shield lichens unique to the volcanic province. Given the diversity of its vegetation, microclimates, and habitat types, Mount Canobolas is likely to be a hotspot for fungi. While local field naturalists have documented the lichen communities, little was known about other groups of fungi. But that was about to change.

Liz is one in a league of rural women driving public awareness of fungi. Her own interest in fungi started as a forager and supplier of saffron milkcaps to Sydney's produce markets. After hearing an orchid biologist reveal the wonders of mycorrhizal relationships, her passion for fungi grew beyond their edibility to include their ecology and conservation. "There's so many layers to the story of fungi we don't understand," Liz told me. "I've loved trying to work them out. It's been the greatest legacy of my career to have this region recognize their importance. There's nothing else I've done that has had the impact of my work with fungi."

The cars that soon arrived on the mount, filled with dozens of survey participants, were testament to her influence. While interest in fungi grows, only the most dedicated mycophiles are willing to tolerate temperatures below zero to crawl around on a windswept mountain in search of them.

One of the first to leap out of her car and pull on her coat and gloves was Jane Paul, now in her tenth decade. Shiny-eyed, with a striped beanie tucked down over her ears, Jane was already off and poking around for fungi before the rest of us had our boots on.

Armed with field guides and collection boxes, the surveyors disappeared into the bush. Over the course of the day, they recorded 40 fungus species on the mount for which there were no previous records. The local rangers promised to include their findings in the mount's management plan. This was a boon for fungi, as they rarely figure in the management plans of national parks or conservation reserves, overshadowed by charismatic mammals, birds, and orchids. Many of these management plans are off-the-shelf documents that contain lists of species. There's usually a list of rare ones for protection, accompanied by a list of "weeds and ferals" for eradication. Only in rare cases are fungi included at all, and then it's usually in the latter list. Park management practices such as fuel-reduction burning are largely modeled around vegetation communities or the specific requirements of endangered animals, not fungi. To have fungi included as a vital part of the biodiversity of a reserve allows for more effective conservation.

As the surveyors discovered the fungi of the mount, I thought about how they were contributing to conservation mostly by crawling on the ground, peering through magnifying glasses, and recording their finds with frozen fingers. It's the long and often uncomfortable slog of environmental survey. But the sense of shared discovery and of contributing to the bigger-picture understanding of fungi sustained them until darkness fell. Conservation efforts to protect species and places from exploitation are often driven by the pluck and mettle of volunteers. And they're usually inspired by one person with a big vision—and in the Central Tablelands of New South Wales, that's Liz Davis. Following the success of the survey, Liz expanded the idea to a month-long calendar of fungus forays, surveys, and events dubbed "Mycology in May." It seems there's no stopping her.

Yolŋu Women's Moon-Shaped Medicine

Liz and mycologists Pam Catcheside, Lynne Boddy, and Suzanne Simard, and radio journalist Amy Mullins, are among the many remarkable women bringing fungi into the public spotlight today. But the oldest keepers of fungal lore in the world could well be Australia's First Nations people. The Yorta Yorta aunties from the Dhungala or Murray River are working to try to retrieve lost knowledge of fungi. In Australia's far north, First Nations people have retained Traditional Knowledge of medicinal fungi and continue to use them for curing illness.

Arnhem Land, in northern Australia, is home to some of the most remote communities of indigenous peoples in the world. The Timor Sea lies to the west and the Arafura Sea to the north. To the east is the great Gulf of Carpentaria, while a vast wilderness of rocky escarpments hiding rainforest remnants lies to the south. The culture of the East Arnhem Land Yolŋu people is among the oldest continuing cultures on earth and remains strong today.

Almost two centuries ago, 36 Royal Marines made what was to be the third unsuccessful attempt by colonists to settle in this region. Little remains to mark their presence today. Park ranger Lyndall McLean and I wandered among the scant stone ruins of the failed Victoria Settlement in West Arnhem Land. Gazing out across the calm and peaceful Arafura Sea, we tried to imagine their lives. We'd sailed in under cloudless skies, accompanied by turtles and the flickering silver bodies of Spanish mackerel. It could not have been more idyllic. But this was not the case for those settlers. Their lives were racked by cyclones, tropical diseases, the threat of crocodiles, and the failure of supply ships to arrive. We bent down to study the hand-etched engravings on the gravestones of those who succumbed to the harsh conditions. It took only eleven years for the settlement to be abandoned. Yet the region had already been inhabited by humans for at least 60,000 years.

I'd met Lyndall while working on a fire research project in Kakadu National Park with the Northern Territory University (now Charles Darwin University). During her 15 years as a ranger she'd spent a lot of time on Country with the Traditional Custodians. I asked her about Indigenous uses of fungi. "I know just the person you should speak with," she said, and put me in touch with videographer Kate Lucas. Kate first went to northern Australia in 1991 to work with the Bininj peoples of West Arnhem Land. She developed a cultural video unit in Kakadu and went on to become an interpretive ranger. More recently, Kate has worked as an adviser to the Goŋ-Däl Aboriginal Corporation in East Arnhem Land, and helped set up a mobile video production unit to record the Traditional Knowledge of elders for future generations.

Working with Kate to document the use of traditional medicines was Yolŋu elder Alison Djawurrku Wunungmurra, from Raymangirr homeland. Through Kate as a proxy, Alison shared her knowledge of a medicinal fungus. "Alison is renowned for her bush medicine knowledge that she learned from her mother," Kate told me. "Families on homelands still use bush medicines to treat common ailments." Kate's video shows Alison removing a bracket fungus from a tree and preparing *wirrmunganing miditjin*—*wirrmunganing* is the name of the fungus, and *miditjin* is Kriol (northern Australian creole) for medicine. Kate tells the story of being on Country with Alison:

> One afternoon we went collecting a plant root used to treat boils, and we ended up collecting several other medicines that were in the area, including the moon-shaped *wirrmunganing* fungus. It was growing on paperbark trees near a saltwater meadow, but Alison said you can get them from other trees in the woodland, although I'm not sure which ones. It's used to treat diarrhea in babies and toddlers. The fungus looks like a half a moon and is placed on the fire until it burns almost to a coal and allowed to cool. It then gets ground into ash and mixed with water to make a paste that is smeared all over the body of the baby. It's left like that overnight and helps take pain and diarrhea away.

Looking at Kate's video, the fungus Alison was using appeared to belong to the genus *Phellinus*. It's a genus of polypores that form hard and tough brackets on the sides of trees.

The Yolŋu people have long and deep connections with some fungi, yet East Arnhem Land is virtually unsurveyed by Western scientists. A specimen of *Phellinus* from Alison's region housed at the Australian National Herbarium in Canberra was identified as *P. rimosus*. The medicinal use of what could have been the same species was previously recorded under the names *mangarnmangarn* and *bulungun/ bulongun*. It's also used for curing sores in the throat by inhaling smoke of the burning fungus, and for relieving coughing by consuming the charred fungus in water. Given various dialects are spoken by the Yolŋu of East Arnhem Land, it's likely that the same fungus could have several names.

A closely related species, *P. robustus*, referred to as *mungun mungun* by the Yolŋu of Galiwinku (Elcho Island), is also used for curing diarrhea with a similar method to Alison's. Another technique involves roasting the fungus on coals, scraping it into water and consuming the mix. While Alison's fungus appeared to be a *Phellinus*, without a specimen, it's hard to be sure which *Phellinus* it was, or whether it was another common polypore such as a *Ganoderma*. However, Alison knew what she was after. I wondered how many generations of women before Alison and her mother had used this fungus, and whether it could be the oldest knowledge of fungi in the world. It struck me as incredible to think of someone sharing knowledge that could be thousands, possibly tens of thousands of years old, and I needed a moment to let that resonate.

It's hard to know the extent to which various Aboriginal groups used fungi. Only a handful of fungus species have been documented as being eaten or used medicinally. The diaries of early explorers, settlers, and missionaries provide some of the earliest written reports of the use of fungi by Aboriginal people. However, records are scant and details usually too limited to be sure of the species used. In his records of his expedition to Central and Western Australia, English explorer Edward Eyre noted in his journals, "Fungi are abundant and of great variety. Some are obtained from the surface of the ground, others below it, and others again from the trunks and boughs of trees." Another English explorer, George Grey, also commented on the abundance of fungi and use by Aboriginal people of the south-west of Western Australia.

Some specimens that were submitted to European herbaria soon after colonization were identified by European mycologists. However, fungi deteriorate quickly and were often in poor condition by the time they traveled by ship across the world. Names were put on specimens based on what they looked like, and after their prolonged journeys they probably didn't look great. Features that were important for identifying fungi could well have been lost. Today, with the benefit of powerful microscopes and molecular technology, identities and names have been revised. Field notes that accompanied specimens occasionally referred to Aboriginal fungus names, but details of how they were used as food or medicine were mostly absent. While these early records tell us something about Aboriginal use of fungi, they are limited. We need a comprehensive effort to work with elders to record, share, and/or revive remaining Indigenous knowledge, with women elders leading the call.

Bringing mycology and Indigenous knowledge of fungi together begins with appreciating the different ways people come to know fungi. I have sought to understand fungi through my own lenses of science, aesthetics, and lived experience, but they are inevitably limited by my own cultural framework. Women's folkloric and mycological knowledge of fungi has played a fundamental role throughout history and continues to do so today. With each interaction and experience, and each sharing of knowledge and ideas, the trove of fungal knowledge is enriched.

10

RESTORING FUNGI

The forest was in a strange mood. It smelled different and was oddly quiet. The soft cries of a pair of yellow-tailed black cockatoos overhead only intensified the silence. Ecologist Amber Sullivan and I were in the High Country of the Victorian Alps searching for aquatic nymphs of the alpine stonefly, a rare and threatened species found only on the Bogong High Plains. Our mission was to document its whereabouts in the hope of averting the clear-felling of the catchments where it lived.

Logging operations and the conservation of rare and threatened species seldom align. Some people might consider a supposedly lowly stonefly an unworthy disruption to logging. However, as an indicator of high-quality habitats, it represented more than *just* a species. The forests around us were some of the most diverse and intact in the state, and likely to harbor a swathe of rare species, including fungi. Weeks spent scrambling through forests of alpine ash, woodlands, and scrublands, searching remote alpine creeks for the nymphs, confirmed its rarity. But in that moment, it was the rarity of birdsong that disturbed us. We clambered across the hillside and saw the creek frothing, the color of milky coffee, in the valley below. It was not a good sign. Only when we rounded the hill did the full extent of the devastation become apparent.

All around us lay the smashed remnants of a broken forest. Crippled limbs lay severed and twisted. The once luxuriant moistness of the forest floor crunched underfoot. And it smelled bad. The usual mingling of forest scents was reduced to the cloying funk of slowly baking mud. Leaf litter and moss that had gently filtered water into the subterrain were gone. Old logs had been bulldozed into piles on the side of the track to allow access to the giant machines of forestry. Along the creek, understory trees were shoved into the buffer zone. This meager strip was supposed to protect waterways from logging disturbance. It clearly failed. Overnight rain had swept the dislocated soils from the steep, exposed slope through the sketchy screen of damaged trees and into the creek. The stonefly nymphs need clean, oxygen-carrying water. They stood little chance of survival in the sediment-laden slurry.

In the middle of the felled area, a solitary tree—a supposed "habitat tree" and a requirement of the logging prescription—stood pathetically alone. Its branches had been wrenched off by the trees that had been felled around it. Little more than a battered and splintered trunk remained. Retaining habitat trees might appear well intentioned, but I realized it was a farce. What animal in its right mind would choose an isolated and damaged habitat tree as home? Trees are not meant to live alone. Now exposed to the elements, its root system riven, and its networks severed, this tree had only a slim chance of survival.

This experience with Amber happened more than three decades ago. It was my first direct exposure to the devastation wrought by clear-felling. I've found myself in dozens of coupes—clear-felled areas—since then, but the heart-crunching pain of that first confrontation is firmly etched in my being. Something profound changed that day. I watched Amber as she stared, glassy-eyed, her nostrils twitching. Neither of us spoke. When you travel along main roads, your view of logging is shielded by carefully planned buffer strips of trees by the roadside. Coupes are mostly unseen. When you travel the backroads, it's a different experience. If you enter a coupe on foot, the transition from intact forest to devastation is acute.

At the time, as a young scientist in my first job, my role was to assess the condition or "health" of river catchments and the ecologies they supported. I knew I was meant to remain detached and objective, and to get on with the job at hand. To think about a river, or a forest, or a fungus *objectively*, as the word suggests, is to regard them as objects. Being objective meant I had to separate myself, the subject, from these observed "objects." But to do so was to create a gap. It meant denying any connection between me and what I observed. Objectivity is at the heart of scientific method. As scientists, we are required to extricate ourselves from the subject of our research. But in that moment, it seemed absurd to try to understand the forest by severing my relationship with it. I wasn't rejecting the objectivity of science, but questioning the measuring of "objects," machine-like, detached from the senses, in isolation from an impassioned body. What if the thing that objectivity took away—that bit that can't be measured or easily articulated—was the most vital part of all?

Recognizing how forests and fungi are affected by disturbance helps us understand how they recover and how we can help restore them. Forests are remarkably resilient, but their capacity to recover depends on the extent of damage and a host of influences. Subterranean supplies of spores and seeds, old logs and stags, and the fine mosaic of nutrient and water stores all help fungi re-establish themselves and ecosystems to recover. But they are more than just "components." Systems-level interactions and processes must also be restored. Because many interactions happen belowground, they are hard to detect. However, keenly observing the nature of undisturbed ecosystems is a good starting point for bringing back fungi in disturbed ones.

Thinking Like a Fungus

Push your hand through the leaf litter and into the soil. Hold it there for a moment. Notice how cool it is a little below the surface. Rub the soil between your fingers. Notice the sensation. How does it feel? Notice what is in it. Can you recognize anything? Any creatures? Any threads of fungi? Notice the texture. Or were you unable to get your fingers below the surface? If you can, take a little soil in your hand. Or if you can't, put your nose directly down to its surface. Take a long deep inhalation. And another. Can you recognize its scents?

As the palette of forest scents becomes familiar, a walk across a disturbed environment, such as an agricultural field, can reveal their absence. A field typically has less biological complexity than a forest, and most fungi— along with their scents—are missing. Fields usually smell more bacterial than fungal. That's because tilling destroys mycelial networks and allows bacteria to dominate. While some scents are absent, others such as fertilizers—which are applied instead of the nutrients naturally provided by fungi—replace them. If heavy machinery is used on the field, its soils can become compacted and lose air spaces. Some anaerobic bacteria produce a "rotten-egg smell" as they release hydrogen sulfide. On the other hand, forest soils have more fungi and their synergies produce more complex scents. Restoring fungi in damaged ecosystems, whether they have been damaged by agriculture or logging, or by natural events like storms or wildfire, begins with being aware of the requirements of fungi. It's about noticing where and how they live, and where they don't. Observing and mimicking natural ecosystems is a good way to start.

Landcare volunteer Sue Brunskill and I lie on the ground and peer at the underside of a log on her property in northeast Victoria. It's riddled with cracks and crevices and holes, some lined with soft mossy upholstery. Sunken cankers and boles accumulate tiny pools of water. Burls and knots and knobby growths track a history of stresses from injury or insects, viruses or fungi, whirling the wood grain into intense patterns. Woodworkers have sought these quirks in wood for centuries, but other organisms have taken advantage of the microtopographies they create for millennia. The log is crisscrossed and squiggled with trails made by beetle larvae. Their cryptic logic is indecipherable, like strange hieroglyphics. Gently lifting a loosened chunk of wood, we peek beneath. It bristles with life. Spineless creatures scamper for cover among an entanglement of spiders' webs, fungus mycelium, and the frass of wood-boring creatures. We replace it and the secret microcosms return to darkness. "Log" seems too small and blunt a word for the bustling activity and industry it conceals.

Sue and I try to imagine where in the log we'd live if we were fungi. Pressing our fingers against it, we notice it is mostly solid, but in places there's a bit of "give." Would it be sufficiently decomposed for our mycelium to course through? And would we colonize the above or underside sections? Where were the coolest, most shaded parts, and where would we be exposed to sun or wind? In the slow decomposition of the log back to soil, a succession of different fungi inhabit it and take part in its dismantling. What other fungal competitors might threaten our existence, and which would happily cohabitate? Ants trek across the log with a cargo of plant fragments, mites cruise around, and we wonder which invertebrates would eat us and which might disperse our spores. It's all just a game, but it's an exercise in looking closely, and imagining the challenges in the life of a fungus in a log.

Thinking about the needs of fungi is about being able to switch scales. By getting close, we discover the various microhabitats inhabited by fungi, and can appreciate why diverse microcosms are important to their colonization. We run our hands over the log and notice how temperature and moisture vary. Its microclimates differ between its suspended parts and those that hug the ground. Discolored sections provide clues as to how water runs across it. A green sheen of algae and moss reveals where it accumulates. We peer at stripes and splotches of color: taupe, beige, liver, and mauve. At first glance, you'd think someone had taken to the wood with a palette of paint, which is why these are known colloquially as paint fungi. Mycologists call them corticioid or resupinate fungi—"resupinate" means inverted, and refers to their spore-bearing surfaces being on their outer sides, rather than tucked underneath as with a mushroom. They're among the most important fungi in the dismantling of the log. Their textures are surprising. Some are waxy or velvety, others powdery or crusty. Some are smooth, others pimpled or intricately wrinkled or warted. The odd one has tiny spines or pores.

Most resupinate fungi are thought to be saprotrophs

(recyclers), but some are mycorrhizal, helping seedlings establish in fallen trees and organic matter. Although they may look benign, the mycelium of some are equipped with a lethal arsenal of specialized spiny structures, adhesive traps, and constrictive rings for snaring and killing nematodes and poaching their nitrogen. Other resupinates spare the nematodes' lives but attach themselves to them, hitching a ride to new terrains.

The activities of different fungi alter both the structure and chemistry of wood. Fungi exude various sugars, acids, and proteins that are used by other organisms. The intimately intertwined lives of fungi and other organisms inhabiting the log are way too complex for us to comprehend by simply looking. But noticing the contours and topographies of the log's microcosms helps us become aware of how variety—in substrates and microclimates—make it suitable for diverse communities of fungi to take up residence.

Fungi Infiltrate Landcare

The Landcare movement in Australia focuses on land restoration, bringing together farmers and conservationists, scientists, and communities to restore damaged environments in both natural and agricultural systems. It's been officially running since 1989, although organized efforts to revegetate damaged land have been happening since the 1950s. Today there are more than 6,000 Landcare and related groups in Australia, and the concept has spread to dozens of other countries. Landcarers tackle all kinds of land degradation issues by regenerating waterways, eradicating weeds and feral animals, controlling erosion, and planting millions of trees. More recently, some have recognized the importance of including fungi in their thinking about land restoration. Sue Brunskill is one of them.

People gravitate to Sue. She's big-hearted, smiles and laughs easily, and imparts her vast knowledge in a way that makes you feel like it's your own. Having grown up on a farm in the Riverina region of New South Wales, Sue is all capability and never shies away from a challenge. She studied horticulture and park management with the intention of becoming a park ranger, but then discovered bush regeneration and found her true calling. In 1992, Sue and her family moved to the tiny settlement of Wooragee in north-east Victoria, and over the last three decades she's worked as a bush regenerator, taught conservation and land management, and thrown herself into Landcare. "Landcare in its various forms is such a positive and practical way to help the environment," Sue explains. "Even on a very small scale like your garden, you can make a big difference to conserving species. Gardens can be fantastic places for attracting wildlife. And the people you meet through Landcare are usually so positive and caring."

Sue tells me she has always found fungi "attractive, interesting, and mysterious." She organizes fungal ecology workshops for the local community, bringing together farmers and Landcarers, Traditional Owners and scientists, actively encouraging them to include fungi in ecosystem restoration. "Fungi are a vital part of our ecosystems," says Sue. "Not just in the breaking down of organic matter and recycling ... I am sure they are involved in so many more aspects of ecosystem health. Sometimes you don't need to know everything about something to realize how important it is!" Having grown up on the land, Sue worries that some people are losing not just a connection with nature but knowledge of how food is produced. "Landcare plays a huge role in helping people understand both natural and farming systems. I think people in every profession need to study ecology. Understanding ecology changes the way we live our lives."

On Sue's property, we wandered around and inspected the trees she and her husband Colin had planted, and their various land restoration projects. We talked about ways to build resilience back into farming landscapes, and how to recreate conditions that foster diverse and vibrant ecologies. For Landcarers accustomed to planting trees, bringing back fungi can be more challenging. Planting trees can provide almost instant satisfaction. It's a rewarding and tangible process of putting seedlings into the soil, caring for them, and watching the visible changes as they grow. Restoring fungi, on the other hand, can be less immediately gratifying. Bringing back fungi is less about planting—or "fungus-ing"—fungi, and more about creating conditions that are favorable to their flourishing.

However, less is known about the needs and interactions of fungi than about plants. There's more guesswork and experimentation involved. Fungi operate over slower time-frames, mostly out of sight in the soil or other organic matter. It's harder to observe or measure the success of restoring fungi, and there isn't the satisfaction of observing the changes in growth that there is with a planted tree. Some fungi don't produce mushrooms for many years. Others don't produce them at all. But if you take a handful of soil and look closely, the fungal threads of hyphae might be apparent.

Sue and the Landcarers understand the value of restoring fungi in ecosystems. They recognize the importance of creating a rich suite of favorable habitats and conditions to support a diversity of fungi. Part of creating ideal conditions for fungi to flourish is keeping an eye on the things that disturb them.

Scales of Disturbance

If you are an ant and a tree limb falls on you, it could cause a radical disturbance to your day. It might cause physical injury, block your path, cut you off from your mates, or, in the worst case, end your life. Likewise, if you are a hairy curtain crust fungus colonizing a limb that someone then throws on their campfire, it is likely to be catastrophic as your mycelium sizzles to death. At the other end of the scale, the violent manifestations of extreme weather amplified by climate change have massive impacts on ecosystems across vast areas.

When a tree is felled, the most obvious effect is that it is no longer there. It is gone. We notice its absence because trees exist—or don't exist—on a scale with which we are familiar. We can see how aboveground ecologies respond to disturbance. However, these aboveground effects reverberate further, and the tumult plays out unseen in the subterrain.

Disturbances to ecosystems result from events or processes that disrupt them, causing changes in the availability of habitats, resources, or microclimates, and the dynamics and relationships between species. Some disturbances occur as part of the natural flux of ecosystems caused by fire, flooding, or when trees are uprooted by wind. These can be regenerative, fostering ecosystem processes and triggering reproduction, such as with pyrophilous fungi that respond to fire.

However, if disturbance is too frequent, intense, prolonged, or widespread, habitats and species are lost. Human disturbances such as forestry, agriculture, and mining have more adverse effects on fungi and forests than natural disturbances. Industrial forestry, for example, manages forests for the sole purpose of high yields of trees, not for their ecology. This means that logging destroys a forest's structure, then homogenizes it. The constant pursuit of increased productivity and efficiency results in forest biomass and diversity being lost, but the effects reach further. Trees capture and store vast amounts of carbon, but when they're felled, carbon dioxide is released into the atmosphere, contributing to global heating. Forests act like natural evaporative air conditioners, directly cooling the planet. They draw water into their leaves from the soil, which then evaporates with a cooling effect. The loss of trees causes the land surface to warm up and local temperatures to rise. Without tree canopies to intercept rain, and root systems that stabilize soils, the ways that water moves through air and soils is also disrupted.

The process of destroying a forest begins when harvesting machinery enters it, long before trees are felled. The sheer weight of the logging machinery compresses and compacts soil, and fungi are less able to penetrate it. Air-filled spaces, once part of the sponge-like nature of the soil, collapse. Subterranean fauna, tree roots, and fungus mycelium are crushed. Fungi, along with almost every other organism that lives in soil, require oxygen to survive. Those that somehow avoid being crushed by the initial compaction will suffocate and die. Most organisms need the water that gently filters through porous soils, but when they're compacted, water can't penetrate and runs off the surface. Without access to water, soil inhabitants wither and die.

When forests are felled, it is not just the organisms themselves that are lost, but the processes they perform. The constant supply of leaves and limbs that provide organic matter is cut off. Soils can no longer be properly recycled or aerated, water cannot be filtered, and other life-giving processes are diminished. Without trees, a new suite of processes comes into play. Rain directly hits the soil and washes away the remaining nutrients. All these changes radically affect fungi. With trees gone and soil disturbed, mycorrhizas are severed and underground lovers torn apart.

To restore an ecosystem, usually you need to respond to the disturbances that have reduced their capacity to flourish. While the succession of plant communities has been studied for over a century, the patterns and mechanisms by which fungi return to damaged ecosystems are less known.

Undisturbed ecosystems comprise fungi from different functional groups or trophic modes—that is, a mix of saprotrophic, mycorrhizal, and parasitic fungi. Each group performs different functions. When ecosystems are disturbed, the ratio of these groups changes; consequently, the dynamics of the system change too. Felling trees severs relationships, and mycorrhizal fungi are lost. As the ecosystem starts to recover, saprotrophic fungi dominate and disturbed soils are restored. Fungi such as the Australian honey fungus are highly adaptable, shifting between parasitic and saprotrophic modes, enabling them to withstand disturbance better than many species. The Australian honey fungus can quickly proliferate, taking advantage of reduced competition from other fungi that are unable to tolerate the disturbance. If you wander through logged ecosystems, you'll notice it's commonly one of the most dominant species. Over time, ecosystems can recover and mycorrhizal species return.

Meeting with a Visionary Mycophile

The benefits of restoring fungi extend beyond the fungi themselves, yet ecologists rarely mention fungi in the restoration of terrestrial ecosystems. But when we recreate the habitats, climates, and conditions that encourage fungi to colonize, ecosystems are rebuilt. Bringing back fungi contributes to architecture in soils, creates habitats for other organisms, and restores ecosystem processes and complexity. As an ecosystem recovers and matures, its diversity and resilience increases.

A couple of hours north-west of Wooragee, in north-east Victoria, Kazuko Yamamoto lives at the foot of Australia's Snowy Mountains. We'd met at a fungus workshop organized by Sue Brunskill a few years earlier, and discussed Kazuko's efforts to restore native fungi on her property. Few people consider fungi in land restoration, and I was delighted to receive her invitation to visit her place.

I drive up her long tree-lined driveway to find her on her verandah, tying bunches of garlic with twine for drying. Baskets of harvested walnuts and lemons sit by the screen door. Kazuko is lithe and strong, and she looks the part in a pink checked shirt beneath denim overalls teamed with an oversize pair of steel-capped work boots. Binoculars and sunhat sit on the table beside her. She is clearly ready for action, but not before greeting me warmly and offering me tea.

Like Sue, Kazuko studied horticulture, and was also a pharmacist. She'd spent most of her life designing gardens on Hokkaido, Japan, and had moved to Australia just over a decade earlier to be with her daughter and granddaughter. Kazuko describes how she'd bought the old 120-acre farm and immediately set about restoring it. She pours me tea then gets up and goes inside the house, returning a moment later with a large wooden box. Placing it carefully on the table, she opens it and half disappears inside, retrieving a photograph that was taken on the day she moved to the property. I look at a flock of forlorn sheep standing listlessly in a sparsely treed landscape as she tells me the story of her restoration work.

Kazuko sifts through maps and aerial photos, planting calendars, vegetation profiles, and sketches of the areas she has restored. In a battered notebook she has lists of fungus, bird, and plant species written in her meticulous handwriting. Its pages are well thumbed from her daily entries and are coming loose from their binding. From the bottom of the box, Kazuko lifts out her herbarium of dried plants, each neatly contained in a wooden flower press beneath pieces of cardboard and parchment paper. And then there are her fungus spore prints—dozens of them, with names, dates, and locations of where she recorded each species written in the top right-hand corner. Most are accompanied by sketches of the mushrooms. It's an impressive account from an astute observer. "I like to keep track of things," says Kazuko. "It's over ten years since the property has been grazed, but it still needs help. I've learned how to restore it by doing it. A little bit each day."

We finish our tea, and she beckons me to follow her through a row of blueberry bushes and walnut trees, past her propagation shed and chicken run, and down into a valley. "This is my teacher," says Kazuko, her arms stretched wide. "By watching this remnant patch of forest, I'm learning how to bring back fungi." Kazuko rises before dawn and begins each day with a long slow walk around the property with her notebook, observing and recording. She tells me how she closely examines fallen wood of different sizes and ages, and the various fungi colonizing it. Generally, areas with the greatest accumulation and diversity of wood have the greatest diversity of fungi. Many fungi have specific requirements, so recreating habitat variety allows for their different preferences. Large, old wood provides some of the most important habitat of all for fungi. It not only has more area, but more specialist habitats. As it ages, its surfaces become more convoluted and furrowed. It shades the ground beneath it and helps retain moisture. Organic matter accumulates around it, providing shelter for invertebrates and small mammals that are vectors for the spores of fungi such as truffles.

Kazuko bends down and picks up a small branch. It's sulfur yellow and covered in a wood-rotting fungus known as golden splash tooth. She takes out her loupe to look at its blunt-toothed surface, "Oh, it's beautiful, just so beautiful," she says, taking her time to examine every inch of it before passing it to me. She then squats by a cluster of rainbow fungi scalloping an old log in bands of gray, purple, and blue. Kazuko tells me she has been watching it for several years.

It's a species long used in traditional Asian medicine, and Kazuko explains how scientists are now trialing its capacity to degrade pesticides and other agricultural chemicals.

The greater the knowledge of how fungi live in undisturbed ecosystems, the greater the chance of successful restoration. Almost all of Kazuko's efforts follow two foundational principles—creating diversity, and minimizing stresses. Making sure that there is diverse organic matter of different species, size, and age increases the range of habitats and microclimates for fungi. A good starting point for minimizing stresses is to survey a site and establish whether the cause of disturbance persists—and if it does, to determine if it will undermine restoration efforts. Reducing or eliminating disturbances that damage or destroy fungus mycelium increases the success of restoration. Digging, tilling, and excavating cause physical disturbances that sever mycelial and mycorrhizal networks. The heavy machinery used in logging radically compacts soils, but even regular vehicles and the hard hooves of stock destroy soil structure and collapse air spaces vital to fungi. Heating soils through fire, overwatering it through irrigation, or poisoning it with chemicals kills or compromises all but the hardiest fungi. Felling trees removes the partners of mycorrhizal fungi. The message is clear: minimize disturbance. As mycologist Tom May succinctly puts it, "Nature is complicated—where the processes are intact, don't mess with them!"

Restoring fungi begins with working with what is already there. It makes more sense to build on existing remnant areas and their biological legacies than to start from scratch. In the longer term, recovery processes are performed by the organisms themselves, but humans can help to establish and catalyze recovery. Kazuko began by surveying the property to determine what biological legacies existed. She found them in the form of stags and old logs, patches of native grasses, and natural depressions in the landscape where water accumulates. The best legacies were the remnant patches of vegetation with mature trees that had more established connections and networks with fungi. These trees have bigger root systems and provide more soil architecture. Some sites might have enough capacity to generate themselves, as long as the stresses are removed. Kazuko determined which areas were most resilient and showed the best prospects for restoration. If soils are not too damaged, and have retained their structure and a bank of seeds, spores, and nutrients, they regenerate themselves, as long as disturbance is minimal. Other soils are too impoverished and need more help.

We cross over a low hill where galahs and sulphur-crested cockatoos, two common native parrot species, scrabble on the ground hunting for seeds and rhizomes. Too busy searching to be bothered by us, they continue feeding. "This area was grazed," says Kazuko. "The black wattles and common cassinia have come back but not much else. Look how bare and compressed the soil is. Soil is not meant to be bare!" I scan the ground and there are a few grasses but also patches that look as if they've been scalded. "In the remnant areas kangaroo grass and tussock grass are thriving, and in spring the everlasting daisies and scaly buttons flower. I want to bring those back here. Luckily, despite the grazing, there's very few weeds." We crouch down and scratch the compressed soil and it's hard to make leeway, but a little further on we see encrusting lichens colonizing the bare ground.

The lichens are part of the cryptogamic crusts that result from intimate associations between fungi, algae, cyano-bacteria, and mosses in the uppermost quarter-inch of the soil. They create and maintain fertility, and fix carbon and nitrogen. By aggregating soil particles, they stabilize them, reducing their loss to wind and water erosion. Although cryptogamic crusts can often tolerate extremes of temperature, drought, and solar radiation, they often go unnoticed and are therefore vulnerable to disturbance from humans and vehicles as well as livestock. Kazuko lies flat on the ground and peers at them through her loupe. "They're like miniature ecosystems, operating like larger ones but on another scale," she observes. "We should be starting with soil crusts rather than trees when restoring ecosystems, especially in drylands. That's what nature does. Soil crusts are the foundation that allow for succession of other species."

A little further on we stop by a fallen eucalypt, although it's not clear where it's fallen from. "My neighbor helped me drag over this candlebark tree that blew over in my driveway and lay it across the contour. It traps organic matter when it rains. He thinks I'm mad and wanted to cut it up for firewood!" says Kazuko, with feisty goodwill. "Cleaning up, cleaning up, cleaning up, that's his aim. He thinks I'm untidy. But it stops erosion and already I've seen fungi in these little islands of leaf litter. It's working, my plan is working!" She is wide-eyed as she claps her palms together. Among the wattles we find the perfect chocolate-brown caps of rooting shanks. Fallen limbs are splashed in the bright orange of the scarlet bracket where the bark is peeling off. As pioneer species, these saprotrophic fungi get to work on recycling organic matter.

The wattles, too, are pioneers, enhancing soil nutrients and fixing atmospheric nitrogen. Some of them are colonized by mistletoe, which provides a rich source of nectar for butterflies and other insects, native birds like honeyeaters, and gliders and possums. When mistletoes die, they create tree hollows that are quickly inhabited by myriad creatures. Their leaves are also rich in phosphorus, nitrogen, and potassium, so they fertilize the soil when they drop. These various fungus and plant pioneers lay the foundations for recovery.

"Everything comes back to habitat protection. Restoring habitat is a good thing, but protecting it in the first place is far better and means less work!" says Kazuko. "Fortunately, the farmer fenced off the creek, so the riparian areas are mostly intact." We push through the understory of prickly teatree shrubs along the creek where an old mountain swamp gum tree has fallen. Tiny yellow spikes of the pretty horn fungus and clusters of the convoluted jelly masses of the white brain fungus tell us that it is in the early stages of decomposition, as these two fungi are also pioneers. We follow the creek line, then climb up over an exposed ridge to an area that Kazuko has revegetated. Below the ridgeline there are clumps of vegetation with small clearings, fringed by native grasses. Honeyeaters dart in and out of the wattles. Nesting boxes have been strung up in the eucalypts. On the ground, a pile of old fallen fence posts—deliberately retained as habitat rather than casually overlooked—is being colonized by lichens.

Kazuko tells me how she'd sourced local seedlings through Landcare but had trouble convincing her neighbor why she'd planted them in a circle. As with everything she does, Kazuko has her logic, which in this instance was to reduce the perimeter, and to minimize the effect of wind and other "edge effects." She carefully chose plants that are adapted to the local climate, soil, and conditions, so are more likely to become established and survive than those from further afield. The seedlings are also likely to contain propagules of fungi that in turn are adapted to the local conditions.

Kazuko tapped into the knowledge of the local community to find out when to plant them, taking advantage of growing conditions and minimizing the effects of weather extremes. "There was just one dead standing tree here when I began, and now you can hardly see it as the planted trees have grown so much," she notes. "I began with a groundcover layer of grasses and understory of shrubs. They were about three-quarters of what I planted. The rest was trees for the overstory." Kazuko adds more plants each year to help create variation in the age of the vegetation, with both younger and more mature plants now flourishing. She makes the most of any rock or fragment of wood, planting seedlings as close as she can so they can benefit from the extra shelter and nutrient supply.

"These patches of secondary forest have some good legacy features," she explains. "I'm trying to trigger their 'ecological memory'—you know, all the structures and species and their interactions that allow the system to rejig itself, to reorganize itself, to recover following its long history of upheaval. Gradually I'm linking the remnants, piecing it back together." Human use of landscapes often separates ecologies and habitats into spatially distinct fragments. Cleared areas or roads isolate the fragments. They usually not only have less available habitat, but are cut off from vital fungal vectors such as digging mammals. Reconnecting remnants is an important part of restoring habitats for fungi.

Bringing Back Fungi

With her focus on maximizing habitat diversity, Kazuko had started to bring back fungi in her decade on the property. It's a slow process, but her patience and her big-picture thinking are paying off. Sometimes people want to speed up the restoration process and "transplant," or "transfungus," by digging up soil or mycelium, or scattering mushrooms. However, spores are not like seeds. Seeds usually have their own food supply, and many can last for extended periods in the soil, as long as they don't get eaten or rot. Some fungi have darkly pigmented resistant spores that are long-lived. However, given the world is not completely overrun by fungi, it appears that most spores quickly lose viability. If a spore doesn't land in a spot with exactly the right combination of conditions, it is likely either to perish or be consumed by invertebrates. Moving soil around also risks moving "unwanted" or pathogenic species that could compromise the restoration.

Fungi grow in particular habitats for a reason. Keeping fungi in their place, or recreating those places, makes better sense than translocating them. While fungi haven't *chosen* their habitats per se, they are usually inextricably bound to them. Separating a fungus from its habitat and moving it elsewhere detaches it from local conditions and the organisms with which it interacts, as well as its history. The place fungi are in is where they flourish. Translocating a fungus would mean knowing how much soil you need, plus which plants, and how many beetles, springtails, old logs, and rainclouds to add to your wheelbarrow. If the fungi are truffles, you might need to throw in a few small marsupials like potoroos or bandicoots as well. Many things come into play to create precisely the right conditions for a translocated fungus to thrive. Instead, retaining and restoring habitat, as Kazuko has done, is probably a better bet for bringing back fungi.

It is difficult to accurately identify many fungi in the field. You usually need a microscope to see the important diagnostic features of sporing bodies. Even with readily recognizable species, such as the fly agaric, molecular studies are showing that what were thought to be single species can be complexes of many similar-looking species, often with only a slight difference in appearance, distribution, or ecology. While taxonomists still use what's called phenotypic information—the observable physical characteristics of a fungus—when describing species, mycology is increasingly a molecular science. The recent rise of molecular-sequencing techniques and metagenomics, aka metabarcoding, has fast-tracked knowledge of the diversity of fungi.

Metabarcoding uses small "barcode" portions of genetic material from an environmental sample, then determines which fungi are present through next-generation DNA and/ or RNA sequencing techniques. The sequences generated are then compared with sequences in online reference libraries. The beauty of this is that cryptic species—those that appear the same as each other but are in fact different—are revealed. Not only are molecular techniques exposing a far greater diversity of species than was previously thought to exist, but they are also advancing understanding of the dynamics and succession of fungal communities. The more we know about the habitats that fungi prefer, and which fungi associate with one another, the more effectively we can restore habitats and conserve fungi in our ecosystems.

The rise of molecular mycology has not only revealed vastly more species, but it has shown how the differences between them are often extremely subtle. This opens up interesting questions not only about the limits (delimitation) of species, but also how precise and accurate we need to be for the task at hand. There's a growing inclination among community conservation groups to collect samples for DNA sequencing. While the revelations can be informative, they're usually difficult to interpret. Sometimes in the excitement of new techniques and the desire to collect data, we can forget to formulate the question that we want the data to answer. Generating huge amounts of data without the ability to derive meaning from them is not usually helpful. Molecular identification does not replace taxonomic expertise. A wander in the field with a mycologist or knowledgeable mycophile usually provides more accurate identification of species than inexperienced people attempting to interpret DNA-based data. The combination of metabarcoding with in situ observation and monitoring provides a more comprehensive picture of fungi than either approach on its own. But when restoring a disturbed ecosystem, improving the quality and diversity of habitat and minimizing stresses remains the best approach.

Untidying

As storm clouds amass on the horizon, we head back to Kazuko's house. There is more tea, this time accompanied by generous servings of cake. I light Kazuko's stoked fire and waited for her to finish transcribing the day's observations into her record book before asking her how she became interested in fungi. She glances at me as if it is an odd question, taking off her glasses to peer at me more closely.

"I'm Japanese! Japanese love mushrooms," she answers, then falters. "Well ... we love our shiitake and our fermented shoyu and sake, but I'm not sure that all Japanese really love fungi. Not in a bigger sense." Kazuko looks wistful and I wonder how she straddles the typically Japanese cultural aesthetic of cultivated simplicity in gardening with her attempts to restore the unwieldy clutter of the Australian bush. She must have read my thoughts.

"Japanese [people] once loved their gardens because they are beautiful, simple, peaceful. And green, very green. We love our mosses in Japan," she laughs. "But today, it's changing. Nature is disappearing from gardens. Many people prefer concrete. Everything must be tidy. But that's not how nature works." Kazuko shakes her head. "I don't understand how people lose their love of nature. People don't like bugs! But they don't realize the same bugs pollinate the food they eat. Gardens were once a source of pride, a connection to nature, but today the social currency has changed." I think about how Kazuko's comments echo Sue's concerns about disconnection from nature and the lack of understanding around food production. Yet Kazuko remains ever-positive. "My daughter tells me I'm out of date. That now we have green walls, not green gardens," she laughs. "So that gives me hope."

Tidy gardens are not just a Japanese phenomenon. The ubiquitous woodchip has become a staple ingredient of gardens in Australia. While they might add some nutrition and help retain moisture, consistently sized and shaped woodchips simplify the heterogeneity of naturally occurring organic matter. In the same way that not all *Homo sapiens* live in identical apartment blocks, not all fungi live in woodchips. This homogenization eliminates fungi that prefer to live in an old log, a sheet of bark or other organic matter that has not been spat out of a machine as a chip of wood. Gardens covered in woodchips reduce natural diversity by minimizing niches, eliminating specialist fungi and allowing generalist fungus species to dominate. Those generalist species are commonly early colonizers, and thus woodchips can retard the natural succession of species. Consequently, these gardens can lose function and resilience.

When an environment is damaged, there's often a sense of urgency to hasten the restoration process. Restorers sometimes look for products to stimulate and enhance regeneration. A broad range of mycorrhizal inoculants— material containing spores or hyphal fragments—is available commercially. I ask Kazuko if she used inoculants in her restoration work. Kazuko shakes her head. "When I worked in horticulture, we used inoculants with orchids, because they need their fungal partners to germinate. They've become important in orchid conservation. But restoring an ecosystem and rebuilding soil biology is more complex than germinating orchids," she said.

Although mycorrhizas are essential to functioning ecosystems, their efficacy as inoculants has come under a lot of scrutiny. While they have been effective in specific situations in horticulture and agriculture, their effectiveness across broad ecosystem types and species is largely untested.

Adding inoculants to a site requires an intimate knowledge of the plant and fungus communities that are present and their receptivity to the inoculant. No matter how much inoculant you throw into an environment, if it's the wrong inoculant for the species present or if the conditions aren't right, it's unlikely to have any benefit. The viability of inoculants is further reduced in the depleted conditions of degraded ecosystems. Although mycorrhizal inoculation can promote plant growth in some specific field situations, rebuilding habitat is a better option.

Caring for Fungi

Kazuko's big-picture vision, along with her attention to every detail, is vital to the success of her work. She is not just bringing back fungi but recreating whole ecosystems. I thought about her meticulous records and documenting, and her friendly resistance to her neighbor's ridicule. I ask her what she considers the most important aspect of her project. Kazuko smiles before answering, "It's simple. It's about caring. I think we all long to care about something deeply. I think we all yearn for a more profound, a more meaningful, a deeper way of being in the world." It isn't the answer I was expecting, and I think about how her careful awareness, her noticing and recording, all seem so scientific but really are about keeping track and letting the ecosystems restore themselves. Rather than controlling or actively managing them, Kazuko is giving them a helping hand. Mostly, though, it is about her personal connection with the landscape. It is about caring.

Justifying the notion of care as a reason for restoring

an ecosystem or conserving biodiversity has always been difficult. It's not just because care doesn't fit neatly into the algorithms or quantifiable models of land managers or funding bodies. Care is also difficult to articulate. It is not an attachment, an afterthought, or an overlay on rational understanding. It is not a box to be ticked, but a genuinely and deeply felt sense. I think about how care is more important than data, and realize how Kazuko's inspiring passion and compassion are more important than any other message she could convey. Care—real care—is endangered.

I ask Kazuko about the Landcarers who had helped her with the restoration, and whether a species on a Red List or under protective legislation attracts more interest. "It's so important that we get species onto Red Lists and conservation agendas," she says. "We need legal protections. But do people care more if a species is listed?" Kazuko pauses to reflect. "Perhaps initially, but in the longer term, probably not. People are mostly involved at an emotional level. They love being in nature. They're fascinated by fungi and find them beautiful. They might not fully understand what they do, but what matters is that they care."

In response to the accelerating degradation of eco-systems globally, the United Nations declared 2021–30 the Decade on Ecosystem Restoration. The declaration is a welcome call to action for governments to urgently accelerate the restoration of ecosystems. A vast magnitude of change is necessary to combat the effects of deforestation, habitat loss, species extinction, and a swathe of issues all compounded by climate heating. Many questions and unknowns need to be resolved before we will know whether a decade of restoration will have a significant effect beyond a declaration of intent.

Governments have consistently failed to meet the global biodiversity targets set by the United Nations more than a decade ago. Natural habitats are destroyed and fragmented, and species continue to disappear. There's also a dearth of convincing ecosystem restoration on which to premise a decade of action. The research for the Sixth Assessment Report (2021–22) of the Intergovernmental Panel on Climate Change underscores the dire state of the climate crisis and provides scientific evidence of the need for urgent action to cool the planet.

At times it's hard not to feel despondent about the lack of political will, and the disparities of wealth and power between countries that continue to hinder progress toward these goals. I clung to a thread of hope as the 2022 national Australian election delivered a new government, reflecting the people's aspirations for increased action on combating climate change. Yet it is people like Sue Brunskill and Kazuko Yamamoto, and the many conservation groups quietly working away in the background, translating ideals and words into actions, who provide real hope. They remind us that we can all do something to nurture, restore, heal, and ensure that the fungal networks and the clandestine liaisons of underground lovers are kept intact.

As I said goodbye to Kazuko, she plied me with a bag of walnuts, the rest of the cake, and a giant pumpkin from her garden. Between us, we wrestled it to the car, laughing and almost falling over, but somehow managed to get it seatbelted into the passenger side. It would be good company for the drive home. Starting the engine, I hesitated. "Kazuko, do you feel optimistic about the future, about fungi and the restoration of Australian ecologies?" I asked.

Kazuko paused for a long time and studied her hands, as if to check that all her fingers were still intact. She then looked up. "I'm not sure how it matters, as my response has little bearing. Or perhaps it's simply the wrong question," she said. "I go through a range of emotions from immense frustration to grief to joy. All are entirely normal. To think of it as a dichotomy of optimism or pessimism is of little use. Glib optimism is no more helpful than despondent pessimism." A trio of kookaburras broke out in a round of laughter from a nearby blackwood. We both smiled, as if reminded not to get overwhelmed by the weight of our environmental predicament. "We must be realistic about the state of things but continue to look after environments with all the care and energy we can muster," Kazuko added. "Think of it as caring for an unwell grandparent. Keep noticing, responding, connecting deeply, giving, loving. But maintain the rage. It's all part of caring." Kazuko put her hand on mine and winked at me, then stood back from the car. I took a deep breath, then drove down her driveway with the pumpkin. In the rear-view mirror I saw Kazuko bent in a long, deep bow. I couldn't be sure if it was a sign of gratitude, or if she'd spotted another mushroom on the lawn.

EPILOGUE

A light rain falls but barely touches the ground. It releases forest scents that linger in the mist. On the forest floor, slugs leave the shelter of their damp retreats and meet in their usual haunts. They're on the prowl, sensing the mushrooms as they push through the earth. The waking fall forest feels alert and alive as the long summer draws to a close.

We huddle around the campfire. A pan full of foraged mushrooms spits and sizzles. We're a mixed group: Wiradjuri, Iranian, Italian, farmer and forager, artist and scientist, newborn baby. We span eight decades and, for one reason or another, fungi draw us together. We share a bottle of wine as tales of fungi weave this way and that, punctuated by laughter, lots of laughter. There are stories of the day's fungus finds told with gestures from saffron-stained hands, and a sense of anticipation for those mushrooms awaiting discovery. Old dreams of foraging rekindle in young minds.

A specimen box is passed around, a fungus in each compartment. Heads bow over with illuminated loupes. There are expressions of delight and surprise, and more laughter. Fungus names are proffered, considered, and contested. The identities of some fungi remain elusive, but new whimsical ones are proposed. The conversation darts here and there. There's word of a new mycelial innovation, and some radical insights, woven with parables stretching back millennia. I gaze around at these shining faces and observe how understanding of fungi coalesces through different voices and approaches: layers of knowledge and knowhow, experience and insight. Curiosity. Except for Leila—the newborn, who sleeps through it all—we're all adults, but I smile and watch how finding fungi reignites the insatiable sense of discovery from childhood.

It's an exciting time to be part of the ever-unfolding story of fungi in ecosystems and human lives. Revelations of subterranean networks and alliances enrich the way we make sense of fungi and how they help ecosystems hang together. They remind us that we're inseparable, that life is often uncertain. It takes time for us to understand fungi. They call for patience, allow us to move nearer to that which language can't grasp. By moving in, looking closer, we might just think further to explore urgent questions about our restless planet. Fungi inspire us to reconfigure our relationship with the natural world and perhaps to reimagine another way into the future.

The astonishing technologies of today allow us to prod and probe, monitor, measure, and decode. Wildfires, avalanches, viruses—all demand our attention. But perhaps the future is more analogue, more authentic, more daring. As mushrooms heave through the earth, they're a reminder that we need to be more disruptive, to agitate and awaken the slumbering rebel. Maybe the future is not about scaling up the present, about more or faster. It might be about transforming the world into feeling.

Black clouds writhe across the range as the air cools in the gathering dusk. My mind drifts back to the fire. A koala bellows a little further west. A dog's bark echoes across the valley. I think about Su and Kazuko, Clément, Eloise, Auntie Greta and Alison, Jenny and Willie, and I try to imagine when and where, and with which mushrooms, we'll next meet. I wonder if Nancy the bettong has abandoned her pajamas to sniff out truffles. The conversation darts from fungus to fungus, subject to subject, diverges then merges like hyphae. I try to keep up, just listening, sitting in reflective silence and focusing on the glowing embers. I'm glad we're all here in the darkness, by the campfire, sharing stories of fungi. I take out my notebook and begin to write.

ACKNOWLEDGMENTS

I owe much to my fellow mycophiles and field companions who crawled with me through the bush, the forest and the desert, and over glaciers and other terrain in search of fungi. I am grateful for their sense of discovery, guidance and assistance, and good humor. Every foray brought new revelations while opening up more questions about fungi and their mysterious ways. I feel incredibly privileged to have experienced so many wild and astonishing places and their fungi in such good company.

Thank you Barbara Thüler, Carmelo Martinez, César Espinoza, Clément Léandre, Denise Morgan, Doug Ralph, Eloise Ferguson, Esther Win Bo, Fred Rhoades, Gisèle Lavigne, Gladys Wilson, Greta Morgan, Hilda Stewart, Ingibjörg Guðmundsdóttir, Isabella Romero, Jan Muir, Jenny Robinson, Lee Whitford, Liz Davis, Lyndall McLean, Magnús Ólafsson, Marco Santostefano, Millie Sutherland Saines, Nancy Bettong, Neil Longmore, Pat Barnes, Rima Truchanas, Sallie Tucker Jones, Sonia Cooper, Steve Trudell, Su Aung, Sue Brunskill, Sue Smith, and Willie Bedford.

I enjoyed inspired and lively conversations with friends and colleagues who generously shared their knowledge and ideas, and helped me wrangle this book into shape. Many read chapter drafts and offered their insights with goodwill and patience. Thank you also to the many scientists, librarians, and researchers who sent me papers and shared their findings with me. Elders and mentors have offered me wise counsel and deepened my thoughts, while those who arrived more recently in this world reminded me that hunting for fungi never fails to bring out the child in us all.

I extend my heartfelt thanks to Alison Cowan, Alison Djawurrku Wunungmurra, Amy Mullins, Anders Dahlberg, Anders Hirell, Bea Harris, Beatrice Senn-Irlet, Bernadette Hince, Bob Edmonds, Bria Floren, Daniel Raudabaugh, David Minter, David Tiller, Diego Bonetto, Frank Graf, Greg Mueller, Giuliana Furci, Guðríður Gyða Eyjólfsdóttir, Heino Lepp, Helen Davey, Ivan Inderbitzin, Jill de Vos, John Walter, Kate Lucas, Kath Lynch, Kathrin Schlup, Katie Syme, Kazuko Yamamoto, Lauren Ré, Lesley Evans, Libby Robin, Lorelei Norvell, Lynda Wilson, Lynne Ainsworth, Lynne Boddy, Marina Lewis, Marion Neumann, Matt Barrett, Michelle Rickerby, Neil Ingram, Nola Birch, Pam Catcheside, Paul Kelly, Penny Algar, Peter Buchanan, Philip Rogosky, Rachel Tham, Ray Palmer, Rob Suisted, Robyn Williams, Roland Stegmann, Sapphire McMullan-Fisher, Scott Redhead, Sequoia Lewien, Sue McClements, Susan Nuske, Sylvie Stegmann, Tarnya Kruger, Terry Evans, Toby Kiers, Tom May, Tony Kelly, Tracey Masterton, and Troy Horn.

I am indebted to the publishing team at NewSouth Publishing, especially Harriet McInerney for the opportunity to bring this book into being; and editor Emma Driver, who patiently queried every word, clarifying and crafting with her wise editorial eye. I thank my brothers, David and Michael, and my father Max for their endless support and encouragement. I'm especially grateful to Valérie Chételat for being my muse, for always listening, and for putting up with mushrooms spore-printing all over the kitchen table. Finally, I extend my gratitude to those who fight to save our forests, woodlands, and other natural environments, and who restore those that have been harmed. Thank you all.

IMAGES

Color illustrations follow page 134.

SPECIES REGISTER

The following is a list of fungi, plants, and animals that appear in this book. Species often have multiple common names. This list refers only to those common names used in the context of this book, with scientific names given on the right. It is ordered alphabetically by common name so readers can easily cross-reference from the names used in the book.

FUNGI

American matsutake	*Tricholoma murrillianum*
angel wings	*Pleurocybella porrigens*
aniseed funnel	*Clitocybe odora*
austral forkgill	*Austropaxillus infundibuliformis*
Australian honey fungus	*Armillaria luteobubalina*
Australian parasol	*Macrolepiota clelandii*
basket fungus	*Ileodictyon cibarium*
beefsteak fungus	*Fistulina hepatica*
birdsnest fungi	family Nidulariaceae
birch polypore	*Fomitopsis betulina*
blackening brittlegill	*Russula nigricans*
black witches' butter	*Exidia glandulosa*
blewit	*Lepista nuda*
blue chanterelle	*Polyozellus atrolazulinus*
blue pinkgill	*Entoloma hochstetteri*
blushing waxcap	*Neohygrocybe ovina*
brittlegill	genus *Russula*
brown puffball	*Bovista nigrescens*
button mushroom	*Agaricus bisporus*
candy cap	*Lactarius rubidus*
cauliflower mushroom	*Sparassis radicata*
chicken of the woods	*Laetiporus sulphureus*

Chinese vegetable caterpillar	*Ophiocordyceps sinensis*
coconut-scented milkcap	*Lactarius glyciosmus*
common puffball	*Lycoperdon perlatum*
coral tooth fungus; *pekepekekiore*	*Hericium coralloides*
crimson waxcap	*Hygrocybe punicea*
curry milkcap	*Lactarius camphoratus*
deathcap	*Amanita phalloides*
deceiver	genus *Laccaria*
dingy waxcap	*Hygrocybe ingrata*
emperor cortinar	*Cortinarius archeri*
ergot	*Claviceps purpurea*
fenugreek milkcap	*Lactarius helvus*
fibrecap	genus *Inocybe*
field mushroom	*Agaricus campestris*
fly agaric	*Amanita muscaria*
ghost fungus	*Omphalotus nidiformis*
golden chanterelle	*Cantharellus cibarius*
golden jelly bells	*Heterotextus peziziformis*
golden splash tooth	*Phlebia subceracea*
gray waxcap	*Cuphophyllus lacmus*
green elfcup	*Chlorociboria aeruginascens*
hairy curtain crust	*Stereum hirsutum*
hakeke; rubber ear	*Auricularia cornea*
handsome club	*Clavulinopsis laeticolor*
harore	*Armillaria novae-zelandiae*
hedgehog mushroom	*Hydnum repandum*
Hexenröhrling; witches' bolete	genera *Suillellus, Neoboletus*
hidden pinkgill	*Entoloma ravinense*
honey fungi	genus *Armillaria*
honey fungus	*Armillaria mellea*
hooded false morel	*Gyromitra infula*
horn of plenty	*Craterellus cornucopioides*
Iceland moss (lichen)	*Cetraria islandica*
king bolete; porcino	*Boletus edulis*
lacquered bracket	*Ganoderma lucidum*
Lane Cove waxcap	*Hygrocybe lanecovensis*
larch bolete	*Suillus grevillei*
leathery sawgill	*Neolentinus dactyloides*
little ping-pong bat	*Panellus pusillus*
lobster mushroom	*Hypomyces lactifluorum*

map lichen	*Rhizocarpon geographicum*
marshmallow bolete	*Fistulinella mollis*
mealy pinkgill	*Entoloma prunuloides*
milkcap	genera *Lactarius, Lactifluus*
milking bonnet	*Mycena galopus*
morel	genus *Morchella*
myrtle rust	*Austropuccinia psidii*
native bread	*Laccocephalum mylittae*
nitrous waxcap	*Neohygrocybe nitrata*
ochre bracket	*Trametes ochracea*
old birch spots	*Arthopyrenia betulicola*
orange mosscap	*Rickenella fibula*
orange ping-pong bat	*Favolaschia claudopus*
Oregon black truffle	*Leucangium carthusianum*
oyster mushroom	*Pleurotus ostreatus*
Pacific golden chanterelle	*Cantharellus formosus*
pagoda fungus	*Podoserpula pusio*
parasol	*Macrolepiota procera*
pekepekekiore; coral tooth fungus	*Hericium coralloides*
Périgord truffle	*Tuber melanosporum*
plantpot dapperling	*Leucocoprinus birnbaumii*
poison fire coral mushroom	*Trichoderma cornu-damae*
poisonpie	*Hebeloma crustuliniforme*
poplar mushroom; *tawaka*	*Cyclocybe parasitica*
porcelain fungus	*Mucidula mucida*
porcino; king bolete	*Boletus edulis*
pretty horn	*Calocera sinensis*
prince, the	*Agaricus augustus*
rainbow fungus	*Trametes versicolor*
red-banded polypore	*Fomitopsis pinicola*
red cage fungus	*Clathrus ruber*
redlead roundhead	*Leratiomyces ceres*
rock-shield lichen	genus *Xanthoparmelia*
rooting shank	*Hymenopellis radicata*
rootshank	*Phaeocollybia benzokauffmanii*
rubber ear; *hakeke*	*Auricularia cornea*
ruby bonnet	*Cruentomycena viscidocruenta*
rufous milkcap	*Lactarius rufus*
saffron milkcap	*Lactarius deliciosus*
sandy stiltball	*Battarrea phalloides*

scarlet bracket	*Pycnoporus coccineus*
scarlet fairy helmet	*Mycena strobilinoidea*
shaggy inkcap	*Coprinus comatus*
skirt webcap	*Cortinarius australiensis*
slippery jack	*Suillus luteus*
speckleberry lichen	genus *Pseudocyphellaria*
spectacular rustgill	*Gymnopilus junonius*
splendid woodwax	*Hygrophorus speciosus*
strawberry bracket	*Aurantiporus pulcherrimus*
stubby brittlegill	*Russula brevipes*
sulphur tuft	*Hypholoma fasciculare*
summer truffle	*Tuber aestivum*
tawaka; poplar mushroom	*Cyclocybe parasitica*
teatree fingers	*Hypocreopsis amplectens*
the prince	*Agaricus augustus*
tinder polypore	*Fomes fomentarius*
variable oysterling	*Crepidotus variabilis*
vegetable caterpillar	*Ophiocordyceps robertsii*
velvet parachute	*Marasmius elegans*
verdigris agaric	*Stropharia aeruginosa*
violet coral	*Clavaria zollingeri*
violet cort	*Cortinarius violaceus*
webcap	genus *Cortinarius*
western elfin saddle	*Helvella verspertina*
white brain	*Tremella fuciformis*
white dyeball	*Pisolithus albus*
white punk	*Laetiporus portentosus*
wine glass fungus	*Podoscypha petalodes*
winter chanterelle	*Craterellus tubaeformis*
wirrmunganing	*Phellinus* sp.*
witches' bolete; *Hexenröhrling*	genera *Suillellus*, *Neoboletus*
witches' cauldron	*Sarcosoma globosum*
witches' hair (lichen)	*Alectoria sarmentosa*
woolly chanterelle	*Turbinellus floccosus*
yellow navel lichen	*Lichenomphalia chromacea*

---

*   Without a specimen, it is not known for certain that *wirrmunganing* is a species of *Phellinus*.

PLANTS

alder	genus *Alnus*
alpine ash	*Eucalyptus delegatensis*
ash	genus *Fraxinus*
beech	*Fagus sylvatica*
big-leaf maple	*Acer macrophyllum*
black wattle	*Acacia mearnsii*
blackwood	*Acacia melanoxylon*
buloke	*Allocasuarina luehmannii*
candlebark	*Eucalyptus rubida*
cat-tail moss	*Isothecium stoloniferum*
cherry ballart	*Exocarpos cupressiformis*
common cassinia	*Cassinia aculeata*
desert stringybark	*Eucalyptus arenacea*
Douglas fir	*Pseudotsuga menziesii*
downy birch	*Betula pubescens*
early Nancy	*Wurmbea dioica*
European oak	*Quercus robur*
everlasting daisy	*Chrysocephalum apiculatum*
grass tree	*Xanthorrhoea australis*
gray moss	*Racomitrium lanuginosum*
kangaroo grass	*Themeda triandra*
licorice fern	*Polypodium glycyrrhiza*
lungwort	genus *Lobaria*
manna gum	*Eucalyptus viminalis*
messmate	*Eucalyptus obliqua*
mountain ash	*Eucalyptus regnans*
mountain hemlock	*Tsuga mertensiana*
mountain swamp gum	*Eucalyptus camphora*
myrtle beech	*Nothofagus cunninghamii*
paperbark	genus *Melaleuca*
paper birch	*Betula papyrifera*
prickly teatree	*Leptospermum continentale*
quandong	*Elaeocarpus angustifolius*
red alder	*Alnus rubra*
river red gum	*Eucalyptus camaldulensis*
scaly buttons	*Leptorhynchos squamatus*
silver fir	*Abies amabilis*
Sitka spruce	*Picea sitchensis*

soursob	*Oxalis pes-caprae*
southern beech	genera *Fuscospora, Lophozonia*
sugar gum	*Eucalyptus cladocalyx*
tabaquillo	*Polylepis australis*
tussock grass	*Poa sieberiana*
vine maple	*Acer circinatum*
western hemlock	*Tsuga heterophylla*
western red cedar	*Thuja plicata*
yellow birch	*Betula alleghaniensis*
yellow box	*Eucalyptus melliodora*
willow	genus *Salix*
windswept spider orchid	*Caladenia dienema*

ANIMALS

alpine chough	*Pyrrhocorax graculus*
alpine stonefly	*Thaumatoperla alpina*
Australasian gannet	*Morus serrator*
banana slug	*Ariolimax columbianus*
broad-shelled turtle	*Chelodina expansa*
brown treecreeper	*Climacteris picumnus*
common ringtail possum	*Pseudocheirus peregrinus*
common snipe	*Gallinago gallinago*
cormorant	family Phalacrocoracidae
eastern bettong	*Bettongia gaimardi*
emu	*Dromaius novaehollandiae*
Eurasian wren	*Troglodytes troglodytes*
fungus gnat	family Keroplatidae
galah	*Eolophus roseicapilla*
gall wasp	family Cynipidae
glowworm	genus *Arachnocampa*
grey shrikethrush; *wititata*	*Colluricincla harmonica*
kookaburra	*Dacelo novaeguineae*
leaf-cutter ant	genera *Atta, Acromyrmex*
night parrot	*Pezoporus occidentalis*
northern bettong	*Bettongia tropica*
numbat	*Myrmecobius fasciatus*
pied currawong	*Strepera graculina*

pied oystercatcher	*Haematopus longirostris*
red-tailed phascogale	*Phascogale calura*
rufous bristlebird	*Dasyornis broadbenti caryochrous*
sand dollar	order Clypeasteroida
scarlet robin	*Petroica boodang*
scuttle fly	family Phoridae
southern red-backed vole	*Myodes gapperi*
southern right whale	*Eubalaena australis*
Spanish mackerel	*Scomberomorus commerson*
sulphur-crested cockatoo	*Cacatua galerita*
superb lyrebird	*Menura novaehollandiae*
swamp wallaby	*Wallabia bicolor*
Victorian swift moth	*Oxycanus diremptus*
white-winged chough	*Corcorax melanorhamphos*
wititata; grey strikethrush	*Colluricincla harmonica*
yellow-footed antechinus	*Antechinus flavipes*
yellow-tailed black cockatoo	*Calyptorhynchus funereus*

GLOSSARY

agaric—a fungus with lamellae (gills) on the underside of the pileus (cap).

Agaricaceae—a large family of fungi within the order Agaricales that includes many familiar mushrooms, such as those of the genera *Agaricus* and *Amanita* with umbrella-shaped sporing bodies, as well as many of the puffballs.

anthropomorphize—to ascribe human attributes to other species.

arthropod—insects and other invertebrates with jointed legs, e.g. arachnids and crustaceans.

bioluminescent—glowing as a result of a biological process.

biomass—the total quantity or weight of organisms in a given area or volume.

bolete—fleshy mushroom with tubes on the underside of the pileus (cap).

bryophytes—plants that reproduce via spores rather than flowers or seeds, e.g. mosses and liverworts.

carcinogen—a substance capable of causing cancer in living tissue.

chitin—a structural polymer that gives fungal walls (and arthropod exoskeletons and mollusk radulas) hardness.

clade— a group of organisms that have evolved from a common ancestor.

Country—in the Australian Aboriginal context, a place or territory as well as the relational and multidimensional notions of belonging and deep connection to land.

coupe—an area of felled trees in a forestry harvesting operation.

crustose—forming or resembling a crust.

cryptic species—species that are similar in appearance but differ in subtle ways (often in their DNA sequences).

cryptogamic crusts—thin layers of living material in the uppermost quarter-inch of soil aggregated by organisms such as fungi, algae, cyanobacteria, and mosses; also called biological soil crusts.

ecocide—destruction of the natural environment by deliberate or negligent human action.

edge effect—changes in population or community structures that occur at the boundary of two or more habitats.

endemic—occurring exclusively in a defined geographic location.

entomopathogenic fungi—those fungi that grow in or on the bodies of insects.

enzyme—protein that catalyzes (speeds up) chemical reactions in living organisms.

epigeous—above ground.

epiphyte—organism that grows on plants but is not parasitic.

epithet—the second part of a species name (starting with a lower-case letter). Often called a "specific epithet."

exoskeleton—a hard, rigid, and protective external body covering found in arthropods such as crustaceans, insects, and spiders.

flagship species—representatives of a group of organisms, or an environment or environmental issue.

forager—someone who seeks wild edible food, including fungi.

forayer—someone who seeks fungi mostly for scientific interest.

fungarium (pl. fungaria)—a reference collection of fungi.

fungus (pl. fungi)—refers to the entire organism, including its mycelium and reproductive structures.

genus (pl. genera)—a group of closely related species.

gill—see *lamella*.

home range—the region in which an animal normally moves about; encompasses all the resources it requires to survive and reproduce.

host range—the number of other species with which a species associates.

hymenium (pl. hymenia)—spore-bearing surface of fungi.

hypha (pl. hyphae)—filament of fungus mycelium.

hypogeous—belowground.

indicator species—a proxy for the state of an ecosystem.

invertebrates—animals without a backbone, such as insects, arachnids, and crustaceans (among many others).

isotope—one of two or more forms of the same chemical element.

IUCN Red List of Threatened Species—a comprehensive information source on the extinction risk of animals, fungi, and plants.

kingdom—in biological classification, the second highest taxonomic rank after domain.

lamella (pl. lamellae)—(also called gill)—vertical plate on underside of pileus (cap) on which the spore-bearing layer is formed.

litter—organic matter such as leaves, twigs, and pieces of bark that accumulate on the ground.

liverwort—a small, non-vascular, flowerless green plant with leaf-like stems or lobed leaves.

loupe—magnifying hand lens.

macrofungi—fungi that produce sporing bodies that are visible to the naked eye such as mushrooms.

microfungi—fungi such as molds and rusts that have microscopic spore-producing structures.

mycelium (pl. mycelia)—the matrix of thread-like branching fungal cells known individually as a hypha (pl. hyphae) that constitute the fungus feeding body.

mycologist—scientist who studies fungi.

mycology—the scientific study of fungi.

mycophagist—fungus eater.

mycophagous—fungus eating.

mycophilia—love of fungi.

mycophobia—fear of fungi.

mycorrhiza—mutualistic symbiotic association between a fungus and the roots of a plant.

mycotoxin—a poisonous substance produced by a fungus.

myriapods—millipedes, centipedes, and kin with bodies made up of numerous similar segments, most of which bear true jointed legs.

parasite—organism living in or on an organism of another species and benefiting from the association.

pathogen—disease-causing agent.

perithecia (sing. **perithecium**)—small flask-shaped sporing bodies.

phenology—the timing of a biological event, such as the appearance of sporing bodies.

pioneer species—hardy species that are the first to colonize bare or barren ecosystems, usually as a result of disturbance.

propagule—a structure that can give rise to a new organism.

pyrophilous—fire loving.

radula—a minutely toothed ribbon-like feeding structure of mollusks used for scraping/rasping food.

Red List—see *IUCN Red List of Threatened Species*.

refugia—havens where organisms can survive during unfavorable conditions.

remnant vegetation—vegetation that has not been significantly disturbed by destructive land use such as logging, agriculture, or fire.

rhizomorphs—fine, ropy aggregates of hyphae.

riparian area—area immediately adjoining a waterway.

ruderal—disturbed by human actions (in reference to habitats).

saprotroph—organism that gains nutrients from dead organic material.

sapwood—young, soft wood, just beneath the bark of a tree trunk, consisting of living tissue.

sclerotium (pl. **sclerotia**)—dormant resting body of some fungi, consisting of a mass of hyphal threads.

spore—basic reproductive unit of a fungus, usually a single cell.

sporing body—spore-bearing structure (also known as "sporophore," "sporocarp" or "fruiting body").

stags—standing dead trees.

stipe—stem or stalk-like structure supporting the pileus (cap) of most mushrooms.

stroma—a mass of fungal tissue with embedded spore-bearing structures.

substrate—substance or material in which a fungus grows and from which it obtains nutrients.

symbiosis—an intimate relationship between two or more different organisms.

taxa (sing. **taxon**)—taxonomic units (e.g. genus) of any rank within a taxonomic system.

taxonomist—a person who names, describes, and classifies organisms.

trophic mode—the way a fungus feeds (also known as "nutritional guild" or fungus "lifestyle").

truffle—a subterranean fungus sporing body.

type specimen—the name-bearer for a species; the original specimen from which a description of a new species is made.

vector—organism that helps another by transporting its reproductive structures.

SELECTED SOURCES

The following source materials are a selection of the many that informed the writing of this book.

Ainsworth G (1976) *Introduction to the History of Mycology*. Cambridge University Press, Cambridge.

Ammitzboll H, Jordan G, Baker S, Freeman J, Bissett A (2021) Diversity and abundance of soil microbial communities decline, and community compositions change with severity of post-logging fire. *Molecular Ecology* 30 (10), 2434–2448.

Arora D (1986) *Mushrooms Demystified*. Ten Speed Press, New York.

Attenborough D (2016) A glowing underground network of fungi. *BBC Two: Attenborough's Life That Glows*. <www.bbc.co.uk/programmes/p03syr6g>, accessed July 27, 2022.

Bennett G (1860) *Gatherings of a Naturalist in Australasia: Being Observations Principally on the Animal and Vegetable Productions of New South Wales, New Zealand, and Some of the Austral Islands*. J. Van Voorst, London.

Boddy L (2001) Fungal community ecology and wood decomposition processes in angiosperms: From standing tree to complete decay of coarse woody debris. *Ecological Bulletins* 49, 43–56.

Boddy L, Hiscox J (2016) Fungal ecology: Principles and mechanisms of colonization and competition by saprotrophic fungi. *Microbiology Spectrum* 4 (6).

Bowd E, Banks S, Strong C, Lindenmayer D (2019) Long-term impacts of wildfire and logging on forest soils. *Nature Geoscience* 12, 113–118.

Brundrett M, Tedersoo T (2018) Evolutionary history of mycorrhizal symbioses and global host plant diversity. *New Phytologist* 220 (4), 1108–1115.

Büntgen U, Egli S, Camarero J, Fischer E, Stobbe U, Kauserud H, Tegel W, Sproll L, Stenseth N (2012) Drought-induced decline in Mediterranean truffle harvest. *Nature Climate Change* 2 (12), 827–829.

Cannon P, Hywel-Jones N, Maczey N, Norbu L, Tshitila, Samdup T, Lhendup P (2009) Steps towards sustainable harvest of *Ophiocordyceps sinensis* in Bhutan. *Biodiversity and Conservation* 18, 2263–2281.

Carrie A, Heegaard E, Høiland K, Senn-Irlet B, Kuyper T, Krisai-Greilhuber I, Kirk P, Heilmann-Clausen J, Gange A, Egli S, et al. (2018) Explaining European fungal fruiting phenology with climate variability. *Ecology* 99 (6), 1306–1315.

Cline E, Ammirati J, Edmonds R (2005) Does proximity to mature trees influence ectomycorrhizal fungus communities of Douglas-fir seedlings? *New Phytologist* 166 (3), 993–1009.

Cooke M C (1895) *Introduction to the Study of Fungi*. Adam and Charles Black, London.

Cronon W (1996) The trouble with wilderness; or getting back to the wrong nature. In: Cronon W, editor. *Uncommon Ground: Rethinking the Human Place in Nature*. Norton & Co., New York, pp. 69–90.

Dahlberg A, Genney D, Heilmann-Clausen J (2010) Developing a comprehensive strategy for fungal conservation in Europe: Current status and future needs. *Fungal Ecology* 3 (2), 50–64.

Dawes M A, Schleppi P, Hättenschwiler S, Rixen C, Hagedorn F (2017) Soil warming opens the nitrogen cycle at the alpine treeline. *Global Change Biology* 23 (1), 421–434.

Diez J, Kauserud H, Andrew C, Heegaard E, Krisai-Greilhuber I, Senn-Irlet B, Høiland K, Egli S, Büntgen U (2020) Altitudinal upwards shifts in fungal fruiting in the Alps. *Proceedings of the Royal Society B*. 287 (1919), 20192348.

Dighton J (2016) *Fungi in Ecosystem Processes*. CRC Press, Boca Raton, FL.

Dove N, Hart S (2017) Fire reduces fungal species richness and in situ mycorrhizal colonization: A meta-analysis. *Fire Ecology* 13, 37–65.

Doyle A C (1906) *Sir Nigel*. Smith Elder and Co., London.

Egli S, Peter M, Buser C, Stahel W, Ayer F (2006) Mushroom picking does not impair future harvests: Results of a long-term study in Switzerland. *Biological Conservation* 129 (2), 271–276.

Eyre, E (1845) *Journals of Expeditions of Discovery into Central Australia, and Overland from Adelaide to King George's Sound, in the Years 1840–1*, vol. 2. Boone, London.

Fine G (1998) *Morel Tales: The Culture of Mushrooming*. Harvard University Press, Cambridge, MA.

Fontaine A S (2018) Icelandic park ranger calls on car rental companies to inform tourists about offroading. *Reykjavik Grapevine*. <https://grapevine.is/news/2018/08/23/icelandic-park-ranger-calls-on-car-rental-companies-to-inform-tourists-about-offroading/>, accessed July 28, 2022.

Fricker M, Heaton L, Jones N, Boddy L (2017). The mycelium as a network. *Microbiology Spectrum* 5 (3).

Fukasawa Y, Matsukura K (2021) Decay stages of wood and associated fungal communities characterise diversity–decomposition relationships. *Scientific Reports* 11, 8972.

Fuller R, Buchanan P, Roberts M (2005) Medicinal uses of fungi by New Zealand Māori people. *International Journal of Medicinal Mushrooms* 7 (3), 398–400.

Gargano M, Venturella G, Ferraro V (2021) Is *Battarrea phalloides* really an endangered species? *Plant Biosystems: An International Journal Dealing with all Aspects of Plant Biology* 155 (4), 759–762.

Gdula A, Konwerski S, Olejniczak I, Rutkowsi T, Skubala P, et al. (2021) The role of bracket fungi in creating alpha diversity of invertebrates in the Białowieża National Park, Poland. *Ecology and Evolution* 11 (11), 6456–6470.

Graf F, Frei M, Böll A (2009) Effects of vegetation on the angle of internal friction of a moraine. *Forest Snow and Landscape Research* 82 (1), 61–77.

Haeckel E (1905) *The Wonders of Life*. McCabe J, translator. Harper, New York.

Havard K, Heegaard E, Semenov M, Boddy L, Halvorsen R, et al. (2010) Climate change and spring-fruiting fungi. *Proceedings of the Royal Society B* 277 (1685), 1169–1177.

Hedlund K, Boddy L, Preston C (1991) Mycelial responses of the soil fungus, *Mortierella isabellina*, to grazing by *Onychiurus armatus* (Collembola). *Soil Biology and Biochemistry* 23 (4), 361–366.

Heilmann-Clausen J, Barron E, Boddy L, Dahlberg A, Griffith G, et al. (2014) A fungal perspective on conservation biology. *Conservation Biology* 29 (1), 61–68.

Kalotas A (1996) Aboriginal knowledge and use of fungi. In: Orchard A, editor. *Fungi of Australia*, vol. 1B. CSIRO Publishing, Melbourne, pp. 269–295.

Kauserud H, Heegaard E, Büntgen U, Halvorsen R, Egli S, et al. (2012) Warming-induced shift in European mushroom fruiting phenology. *Proceedings of the National Academy of Sciences of the United States of America* 109 (36), 14488–14493.

Kearney S, Carwardine J, Reside A, Adams V, Nelson R, et al. (2022) Saving species beyond the protected area fence: Threats must be managed across multiple land tenure types to secure Australia's endangered species. *Conservation Science and Practice*, 4 (3), e617.

Keewaydinoquay (1998) *Puhpohwee for the People: A Narrative Account of Some Uses of Fungi Among the Ahnishinaabeg*. Northern Illinois University Press, DeKalb, IL.

Kendrick B (1992) *The Fifth Kingdom*. Mycologue Publications, Newburyport, MA.

Kim H N, Do H H, Seo J S, Kim H Y (2016) Two cases of incidental *Podostroma cornu-damae* poisoning. *Clinical and Experimental Emergency Medicine* 3 (3), 186–189.

Kimmerer R W (2013) *Braiding Sweetgrass*. Milkweed Editions, Minneapolis, MN.

Krishnamurthy K, Francis R (2012) A critical review on the utility of DNA barcoding in biodiversity conservation. *Biodiversity Conservation* 21, 1901–1919.

Laperriere G, Desgagné-Penix I, Germain H (2018) DNA distribution pattern and metabolite profile of wild edible lobster mushroom (*Hypomyces lactifluorum/Russula brevipes*). *Genome* 61 (5), 329–336.

Lindenmayer D, Laurance W, Franklin J (2012) Global decline in large old trees. *Science* 338, 1305–1306.

Lower P, Pike J (1990) *Jilji: Life in the Great Sandy Desert*. Magabala Books, Broome.

Marley G (2010) *Chanterelle Dreams, Amanita Nightmares*. Chelsea Green Publishing, White River Junction, VT.

Maroske S, May T, Taylor A, Vaughan A, Lucas A (2018) On the threshold of mycology: Flora Martin née Campbell (1845–1923). *Muelleria* 36, 51–73.

Maser C, Claridge A, Trappe J (2010) *Trees, Truffles, and Beasts: How Forests Function*. Rutgers University Press, New Brunswick, NJ.

Matheny P, Swenie R, Miller A, Petersen R, Hughes K (2018) Revision of pyrophilous taxa of *Pholiota* described from North America reveals four species—*P. brunnescens*, *P. castanea*, *P. highlandensis*, and *P. molesta*. *Mycologia* 110 (6), 997–1016.

May T (2001) Documenting the fungal biodiversity of Australasia: From 1800 to 2000 and beyond. *Australian Systematic Botany* 14 (3), 329–356.

———, Cooper J, Dahlberg A, Furci G, Minter D, et al. (2018) Recognition of the discipline of conservation mycology. *Conservation Biology* 33, 73–76.

McCoy P (2016) *Radical Mycology*. Chthaeus Press, Portland, OR.

Mir-Tutusaus J, Masís-Mora M, Corcellas C, Eljarrat E, Barceló D, et al. (2014) Degradation of selected agrochemicals by the white rot fungus *Trametes versicolor*. *Science of the Total Environment* 1 (500–501), 235–242.

Money N (2006) *The Triumph of the Fungi: A Rotten History*. Oxford University Press, Oxford.

Moore D (1998) *Fungal Morphogenesis*. Cambridge University Press, Cambridge.

——— (2013) *Fungal Biology in the Origin and Emergence of Life*. Cambridge University Press, Cambridge.

———, Robson G, Trinci A (2011) *21st Century Guidebook to Fungi*. Cambridge University Press, Cambridge.

Moos C, Bebi P, Graf F, Mattli J, Rickli C, Schwarz M (2016) How does forest structure affect root reinforcement and susceptibility to shallow landslides? *Earth Surface Processes and Landforms* 41 (7), 951–960.

Nees von Esenbeck T (1823) Correspondenz. *Flora* 6, 115–123.

Norvell L (1995) Loving the chanterelle to death? The ten-year Oregon Chanterelle Project. *McIlvainea: Journal of American Amateur Mycology* 12 (1), 6–25.

Nuske S, Anslan S, Tedersoo L, Bonner M, Congdon B, Abell S (2018) The endangered northern bettong, *Bettongia tropica*, performs a unique and potentially irreplaceable dispersal function for ectomycorrhizal truffle fungi. *Molecular Ecology* 27 (23), 4960–4971.

Ori F, Menotta M, Leonardi M, Amicucci A, Zambonelli A, et al. (2021) Effect of slug mycophagy on *Tuber aestivum* spores. *Fungal Biology* 125 (10), 796–805.

Parfitt D, Hunt J, Dockrell D, Rogers H, Boddy L (2010) Do all trees carry the seed of their own destruction? PCR reveals numerous wood decay fungi latently present in sapwood of a wide range of angiosperm trees. *Fungal Ecology* 3 (4), 338–346.

Phillips M (2017) *Mycorrhizal Planet*. Chelsea Green Publishing, White River Junction, VT.

Phillips R (2006) *Mushrooms*. Pan Macmillan, London.

Pouliot A (2018) *The Allure of Fungi*. CSIRO Publishing, Melbourne.

———, May T (2011) The third "F"—fungi in Australian biodiversity conservation: Actions, issues and initiatives. *Mycologica Balcanica* 7, 41–48.

———, May T (2021) *Wild Mushrooming: A Guide for Foragers*. CSIRO Publishing, Melbourne.

Pyne S (2019) The planet is burning. *Aeon*. <www.aeon.co/essays/the-planet-is-burning-around-us-is-it-time-to-declare-the-pyrocene>, accessed December 3, 2021.

——— (2021) The Pyrocene. *Stephen J. Pyne*. <www.stephenpyne.com/disc.htm>, accessed January 9, 2022.

Raudabaugh D, Matheny B, Hughes K, Iturriaga T, Sargent M, Miller A (2020) Where are they hiding? Testing the body snatchers hypothesis in pyrophilous fungi. *Fungal Ecology* 43, 100870.

Rayner A (1997) *Degrees of Freedom: Living in Dynamic Boundaries*. Imperial College Press, London.

Robin L (2007) *How a Continent Created a Nation*. UNSW Press, Sydney.

Sapp J (1994) *Evolution by Association: A History of Symbioses*. Oxford University Press, New York.

Schwilk D, Keeley J, Bond W (1997) The intermediate disturbance hypothesis does not explain fire and diversity pattern in fynbos. *Plant Ecology* 132, 77–84.

Seidl R, Rammer W, Spies T (2014) Disturbance legacies increase the resilience of forest ecosystem structure, composition, and functioning. *Ecological Applications* 24 (8), 2063–2077.

Sheldrake M (2020) *Entangled Life*. Bodley Head, London.

Siitonen S (2001) Forest management, coarse woody debris and saproxylic organisms: Fennoscandian boreal forests as an example. *Ecological Bulletins* 49, 11–41.

Simard S (2021a) *Finding the Mother Tree*. Penguin Random House, London.

Solly E, Lindahl B, Dawes M, Peter M, Souza R, et al. (2017) Experimental soil warming shifts the fungal community composition at the alpine treeline. *New Phytologist* 215 (2), 766–778.

Steffensen V (2020) *Fire Country: How Indigenous Fire Management Could Help Save Australia*. Hardie Grant Publishing, Melbourne.

Stolt S (n.d.) WA State personal use mushroom harvesting rules. *Puget Sound Mycological Society*. <www.psms.org/WAMushroomRulesMay2016.pdf>, accessed July 9, 2022.

Tedersoo L, May T, Smith M (2010) Ectomycorrhizal lifestyle in fungi: Global diversity, distribution, and evolution of phylogenetic lineages. *Mycorrhiza* 20 (4), 217–263.

Tippett K (host) (2021) Suzanne Simard: Forests are wired for wisdom. *On Being with Krista Tippett* (podcast). <www.onbeing.org/programs/suzanne-simard-forests-are-wired-for-wisdom/>, accessed March 2, 2022.

Tordoff G, Boddy L, Jones H (2006) Grazing by *Folsomia candida* (Collembola) differentially affects mycelial morphology of the cord-forming basidiomycetes *Hypholoma fasciculare*, *Phanerochaete velutina* and *Resinicium bicolor*. *Mycological Research* 110 (e 3), 335–345.

Trappe J, Claridge A, Claridge D, Liddle L (2008) Desert truffles of the Australian outback: Ecology, ethnomycology and taxonomy. *Economic Botany* 62 (3), 497–506.

Vellinga E, Wolfe B, Pringle A (2009) Global patterns of ectomycorrhizal introductions. *New Phytologist* 181 (4), 960–973.

Vitasse Y, Ursenbacher S, Klein G, Bohnenstengel T, Chittaro Y, et al. (2021) Phenological and elevational shifts of plants, animals and fungi under climate change in the European Alps. *Biological Reviews* 96 (5), 1816–1835.

Weinstein P, Delean S, Wood T, Austin A (2016) Bioluminescence in the ghost fungus *Omphalotus nidiformis* does not attract potential spore dispersing insects. *IMA Fungus* 7 (2), 229–234.

Yamin-Pasternak S (2007) How the devils went deaf: Ethnomycology, cuisine and perception of landscape in the Russian north [PhD thesis]. University of Alaska Fairbanks, Fairbanks, AK.

Zhang Q-Y, Dai Y-C (2021) Taxonomy and phylogeny of the *Favolaschia calocera* complex (Mycenaceae) with descriptions of four new species. *Forests* 12 (10), 1397.

INDEX